BRITISH LAND SNAILS

A NEW SERIES
Synopses of the British Fauna
No. 6

BRITISH LAND SNAILS
MOLLUSCA: GASTROPODA
Keys and Notes for the Identification of the Species

R. A. D. CAMERON

AND

MARGARET REDFERN

*Department of Extramural Studies,
The University of Birmingham,
P.O. Box 363,
Birmingham B15 2TT*

1976
Published for
THE LINNEAN SOCIETY OF LONDON
by
ACADEMIC PRESS
LONDON AND NEW YORK

ACADEMIC PRESS INC. (LONDON) LTD
24–28 Oval Road
London, NW1 7DX

U.S. Edition published by
ACADEMIC PRESS INC.
111 Fifth Avenue
New York, New York 10003

Copyright © 1976 by the Linnean Society of London

All Rights Reserved

No part of this book may be reproduced in any form by photostat, microfilm, or any other means, without written permission from the publishers

Library of Congress Catalog Card Number: 75–46327
ISBN: 0–12–157050–9

Text set in 9/10 pt. Monotype Times New Roman, printed by photolithography, and bound in Great Britain at The Pitman Press, Bath

Foreword

British Land Snails is No. 6 of the *New Series* of the *Synopses of the British Fauna* published by Academic Press for the Linnean Society. The *Synopses* are working field and laboratory pocket-books designed to meet the needs of keen amateur naturalists, students and professional biologists, who require a more detailed account of a group of animals than that given in the popular general field-guides. Spaces have been left in the text for the owner's notes, though unfortunately in the interests of economy and size, these spaces are not uniform and often very small. As with previous *Synopses* the cover is waterproofed.

British Land Snails should be of interest to parasitologists as well as field naturalists as snails are often the hosts to many helminth and other parasites. As with the *Synopses*, *British Land Snails* is written by specialists and the Linnean Society is grateful to the authors for the time and care they have given in the preparation of this *Synopsis*.

The Linnean Society is anxious to extend the range of titles in *New Series* of *Synopses* and a number of authors have already promised manuscripts. The Society welcomes both suggestions and offers of new titles, along with comments, so that the series maintains a high standard of content coupled with the maximum degree of usefulness in the field and laboratory.

DORIS M. KERMACK
Synopses Editor, Linnean Society.

A Synopsis of the British Land Snails (Mollusca : Gastropoda)

R. A. D. CAMERON
AND
MARGARET REDFERN

Department of Extramural Studies, University of Birmingham, Birmingham, England

CONTENTS

	Page
Introduction	1
Structure, based on stylommatophoran pulmonates	2
General Biology	6
Collecting and preservation	9
Classification	10
Key to the British species	13
Systematic Part	22
Acknowledgements	61
References	62
Index of species	63

Introduction

British land snails belong to two subclasses of the class Gastropoda in the phylum Mollusca. Two species are in subclass Prosobranchia; all the rest are in two orders in the subclass Pulmonata. Omitting introduced species confined to greenhouses, there are 87 native and naturalized species of land snails in the British Isles, two of which, *Vitrina pyrenaica* and *Vertigo geyeri* are restricted to Ireland. All 87 species are included in this synopsis.

The synopsis does not include land slugs—members of the pulmonate families Testacellidae, Arionidae and Limacidae. These may be immediately distinguished from snails either by lacking an external shell (Arionidae and Limacidae) or by having a very small shell posteriorly into which the animal cannot possibly withdraw (Testacellidae).

The slug families are undergoing considerable revision at species level, as yet not fully published. Details of diagnostic features, ecology and distribution are consequently not available for all species. Quick (1949), supplemented by Ellis (1969) summarize existing information.

Freshwater snails and those found at or below high tide mark are also excluded. When collecting from marginal habitats—near the sea shore, on river banks or in wetlands, Macan (1969) and Ellis (1969) should be used in conjunction with this synopsis.

With some experience, most species of land snail can be identified by shell characters alone. External features of the body provide extra characters, and dissection is necessary only in a few cases. Hence, the key in this synopsis is based, as far as possible, on external characters. The classification of land snails, however, is based mainly on the internal anatomy, particularly that of the reproductive system. The key, therefore, does not identify families; it is a single dichotomous key terminating in a species at each end point. Species keying out close together are not necessarily closely related.

Structure
(based mainly on stylommatophoran pulmonates)

External Features

The *bodies* of nearly all land-snails can be withdrawn completely into their *shells*, but when actively crawling, the main features of the body can be seen (Figure 1a).

The *head* is the anterior part of the body, carrying two pairs of *tentacles*. The posterior pair are above, and larger than, the anterior, and they bear an *eyespot* at the tip of each in all the Stylommatophora except *Cecilioides acicula*, which is blind. The *mouth* lies ventrally between the anterior tentacles.

The *foot* of all the Pulmonata, and of the prosobranch *Acicula fusca*, is simple and undivided, and the animal crawls by muscular waves passing forward along the foot. In the other prosobranch, *Pomatias elegans*, the foot is split longitudinally, and the animal walks by moving each half forward in turn. Only the prosobranchs have an *operculum*, carried dorsally behind the shell when the snail is crawling.

The *visceral mass*, containing most of the digestive and reproductive organs, is contained in the shell. Apart from the connection of foot and head with the rest of the body, the shell mouth is filled with the external wall of the *mantle cavity*, which acts as a lung. There is a *pore* or *pneumostome* which opens and closes periodically to let air in and out of the cavity.

When the snail is withdrawn into its shell, the foot and head are usually carried behind the external wall of the *mantle*, which has considerable powers of water retention. Frequently, a mucous film (the *epiphragm*) is secreted over the mouth of the shell. This is usually thin, transparent and fragile, but some of the larger snails secrete a thicker, opaque epiphragm during hibernation. A succession of epiphragms may be found inside one shell as the inactive snail loses weight and shrinks back up the shell. The prosobranchs seal the mouth with the operculum when inactive.

FIG. 1. A snail and several shells illustrating characters used for identification. (a) *Helix hortensis*, crawling; (b) dextral (*Vertigo substriata*) and sinistral (*Vertigo pusilla*) shells; (c) *Helicigona lapicida*, side view; (d) *Helicella virgata*, side and bottom views; (e) top views of shells with slowly expanding whorls (*Vitrea contracta*, left) and rapidly expanding whorls (*Retinella radiatula*, right).

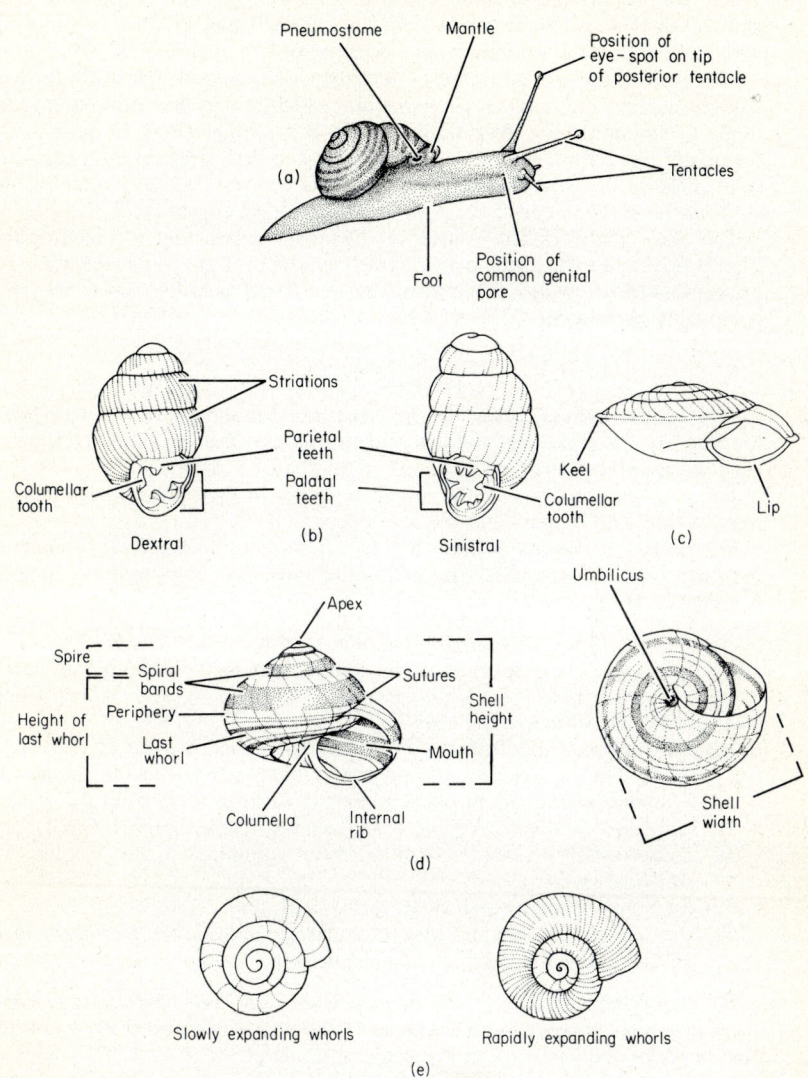

The *shells* of all British land snails are spirally coiled. The direction of coiling sinistral or dextral (Figure 1b) is constant for each species, although specimens with reversed coiling are found very occasionally.

The shell has two major layers—the inner or *ostracum*, relatively thick and calcareous, and the outer or *periostracum*, which is thin and transparent, being made of the chitin-like conchiolin. Bands and blotches may be found in the ostracum, but both layers may be coloured. The periostracum disappears from dead shells in the wild, and may be badly eroded even on living animals. Shells without the periostracum have a whiter and rougher appearance.

The shell is the richest source of diagnostic characters for identification. Figure 1c, d and e shows the main characters used in the key and systematic part. Measurements of width are always the maximum possible, since no shell is completely circular.

Internal Features

Figure 2 shows a dissection of the reproductive tract and gut of *Helix aspersa*. Of the internal features, only the jaw and radula and the reproductive tracts are used for identification.

The *jaw* lies dorsally and the *radula* ventrally in the mouth. The *alimentary canal* is long and looped, and the *anus* opens into the mantle cavity.

Pulmonates are hermaphrodite, and in stylommatophorans there is a common reproductive opening on the right side of the body, opening into the atrium. This is joined in *H. aspersa* by

(i) the evertible *penis*, in a muscular sheath, which tapers down to an *epiphallus*. This splits into the long, blind *flagellum* and into the *vas deferens*, which is separate for a part of its length, after which it lies in close apposition to the *oviduct*;

(ii) the *vagina*, which leads into the convoluted oviduct up to the *albumen gland*. The vas deferens lies on the oviduct, and both enter the gland. A common *genital duct* passes the gametes of both sexes from the *ovotestis*, embedded in digestive gland near the apex, to the albumen gland;

(iii) a duct which divides into a blind diverticulum and a duct leading to the spherical *spermatheca;*

(iv) the many-branched pair of *mucous glands*, and

(v) the *dart-sac*, a large and very muscular sac holding the *love-dart* which is discharged at the partner during courtship.

The reproductive tract of *H. aspersa* is one of the most complex; some accessory parts of this system are absent in some species, and there is considerable inter-specific variation in the size and shape of the organs. The shape and size of the dart may also be diagnostic.

During mating each snail inserts its penis into the vagina of its partner, and transfers sperm in a long spermatophore formed in the flagellum; this is passed by the recipient into the diverticulum of the bursa duct, or spermatheca, and is eventually broken down releasing the sperm. The sperm may be stored in the bursa itself, and later released for fertilization.

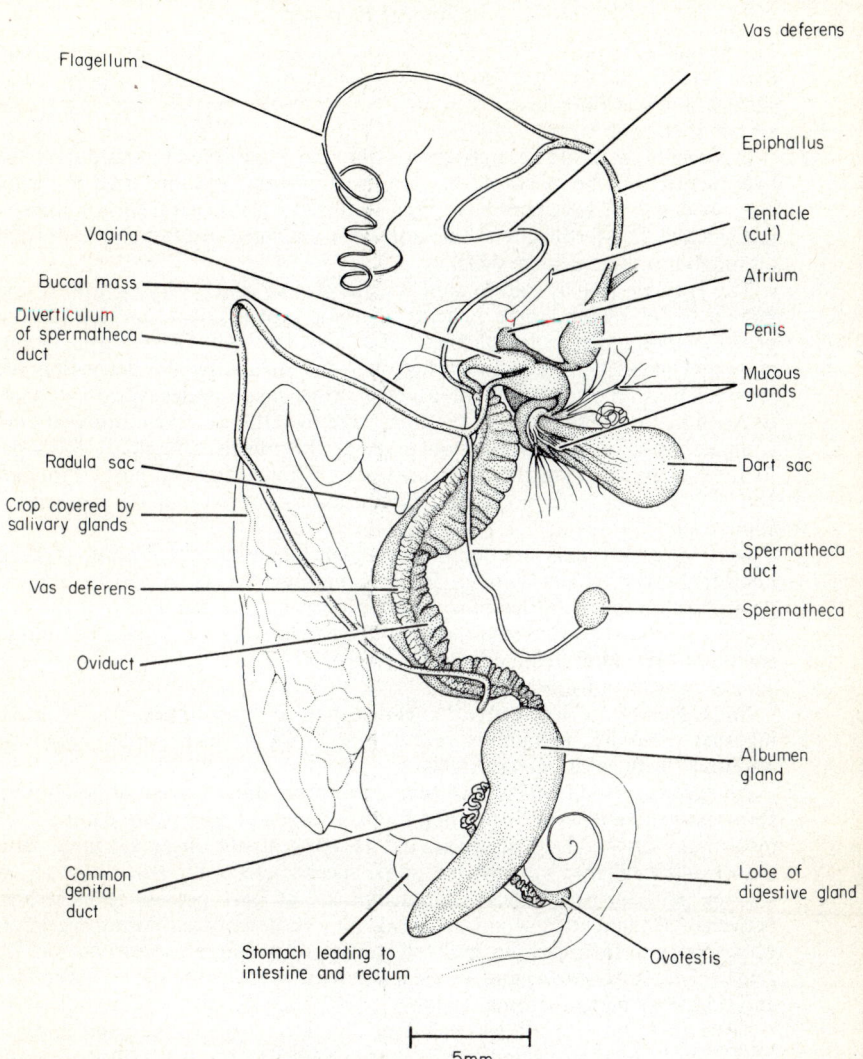

Fig. 2. Dissection of the reproductive system of *Helix aspersa*.

General Biology

Land snails are world-wide in distribution, pulmonates in particular being found from polar regions to the equator. In general, they are inhabitants of the soil surface and the litter layer, although a few species burrow, and some climb several metres up trees.

Pulmonate snails are hermaphrodite, and may mate several times during their lives. Sperm may be stored in the recipient for years, and then used to fertilize eggs. A clutch of eggs may have been fertilized with sperm from a number of individuals. Self-fertilization has not been recorded in British land snails, although it occurs in some slugs.

The eggs are usually laid in clutches, the size of clutch varying with species, but up to 200 in the largest helicids. Development is direct, and the juveniles grow by adding material to the leading edge of the shell. Growth usually stops at sexual maturity, which is marked in some species by the deposition of a strengthening lip or rib at the mouth-edge. Most small species take about a year to reach maturity, but large species may take two, three or even four years.

Snails move on a muscular foot coated with mucus. Muscular waves pass forward along the foot, the contracted parts being raised and pushed forward. Mucus is secreted continuously, so that the passage of the animal is marked by a slime-trail.

Most activity occurs in conditions of high humidity; hence most snails are nocturnal except in wet weather. In dry conditions they cease activity and may deposit a thin sheet of dried mucus, the epiphragm, over the mouth of the shell. If dry conditions persist, respiration and water loss are much reduced, and the snail aestivates until damp conditions return. The larger species can survive for several months in these conditions.

Most species are less active in winter than at other times. A few species hibernate, burying themselves and laying down a thick winter epiphragm reinforced with calcareous granules.

All species of snail which have been studied are polyphagous, although some show marked preferences for certain foods. Living and dead plant material is the most important food. Most species eat decaying plants, algae and fungi rather than healthy "higher" plants, but some species, especially *Helix aspersa*, can become pests on garden plants. Some species, at least, possess cellulases, and between 30% and 70% of natural foods may be assimilated. Many species are very opportunistic in feeding, and will congregate round refuse and carrion. The genera *Oxychilus*, *Vitrina* and *Vitrea* are at least partly carnivorous, other snails and their eggs being the usual victims.

Little quantitative work has been done on the ecology of snails, and Boycott's (1934) review is still the standard work. Dampness and the availability of calcium seem to be the most important factors affecting the local distributions of snails. Active snails have poor powers of water conservation, and calcium is necessary for shell formation. In very general terms, one may recognize three major habitat types, each of which has a characteristic type of snail fauna.

(i) Sand-dunes and calcareous grasslands—liable to intense insolation and drying.
(ii) Woodlands not liable to extremes of drying or waterlogging.
(iii) Wetlands—fens, carrs and marshes.

Most man-made habitats and intermediate zones, such as scrub, have mixtures of species found in two or three of the major types. Some species have very broad ranges of habitat, and may be found in all three of the above types.

Local climate affects faunas markedly, and some very rainy areas may have grassland faunas consisting almost entirely of "woodland" species. Reduction in available calcium in any habitat reduces both the number of species and the average population densities, although only a few species are strictly calcicole. Very acid grassland, peat bogs and heather moors are frequently snailless. Acid woodlands, however, may have a rich fauna if damp and undisturbed.

A number of species are affected by other factors, the most important of which are:
(i) temperature—which appears to determine the geographical distribution of some species.
(ii) rocks and screes—which provide a habitat for some species whether in grassland or woodland.
(iii) the activities of man—which influence many species. A few species are anthropophobes, being intolerant of any disturbance. Others have benefited and are common in gardens and farmland.

Geological and archeological evidence suggests that a few species, now quite widespread and not necessarily associated with man, were introduced in late Iron Age, Roman, or medieval times. The most obvious example is *Helix aspersa*, the garden snail. A few others, introduced more recently, have very restricted distributions.

Causes of death are not well known. Losses in early life are heavy, and may often be caused by adverse climate. Amongst vertebrate predators, the song thrush *Turdus philomelos* is the best known, from its habit of breaking the shells of large species in a characteristic way on stones, leaving piles of shells round the "anvils". Many other birds also eat snails, swallowing small ones whole, and smashing larger ones with their beaks. Shrews, hedgehogs and mice also eat snails.

Amongst invertebrates, several beetles prey on snails, especially the glowworm *Lampyris noctiluca*. Diptera of the family Sciomyzidae are parasitic on slugs and snails, but the extent of their operations is unknown. Carnivorous snails can be important egg predators. Many of the smaller species seem to die off shortly after breeding, but the larger species can live for many years individuals at least 8 years old have been recorded in the wild.

Many species of snail have highly variable shells. Size and shape vary within and between populations of the same species, and this continuous variation is at least partly inherited. Some variants such as sinistrally-coiled specimens of normally dextral species, and monstrosities—gross distortions of the normal shape—are rare and not always inherited. Variation in the colour of the shell, and of the various bands and blotches on it, is usually discontinuous and entirely inherited. Since many such variants may be found in a single population, they are examples of stable genetic polymorphisms, and they have been the subject of many studies. The colour and banding polymorphisms of *Helix nemoralis* (L) and *Helix hortensis* (Müller) have been the most investigated, but many other species show similar, if less dramatic variation. Variation in the frequencies of the various forms between populations can sometimes be ascribed to the action of natural selection for crypsis or resistance to climatic extremes, but the mechanisms maintaining the polymorphisms have not yet been discovered. Ford (1964) gives a useful outline. In a number of species variation is restricted to the occurrence of albino-shelled individuals in an otherwise normal population.

Collecting and Preservation

Snails may be collected by hand in any habitat. Since most species are nocturnal and damp-loving, it is necessary to turn over logs, stones or litter to find them in dry weather. Logs and stones should, of course, be replaced in their original position after the search. Even if done carefully, this method is biased against the smaller species. Leaf-litter and soil may be brought back and hand-sorted in the laboratory. Passing the litter through graded soil sieves down to 0·5 mm mesh, and examining each fraction separately speeds this up. In long grass or fens, a tray may be placed underneath tussocks and the vegetation over it beaten with a stick. For quantitative work, more elaborate methods are necessary.

With practice, many of the larger species can be identified on sight, and it is unnecessary to collect large numbers of living animals. In calcareous sites, many empty shells may be found in the litter and soil if required for large collections or educational purposes.

Living snails are best kept in dry, non airtight boxes pending examination. Nearly all species will survive a few days of dryness. Large snails can eat their way through flimsy cardboard if this gets damp and they will also eat labels, which should always be attached to the outside of the box. If slugs are collected at the same time, they should be kept damp in moist leaf litter if required alive, and are best separated immediately from the snails.

For most species, the shell alone is required for identification. The body is most easily removed by immersing the whole animal in boiling water for a few minutes and then extracting it with a bent pin. This is not always completely successful, and is most difficult with small and fragile species. An alternative is to kill the snail by boiling and let the body rot in the shell in a damp container until it may be dislodged by a jet of water directed into the shell mouth. For very small species, it is not worth attempting to remove the body, which may be left to dry out in the shell. Shells may be stored dry; colours fade if exposed continuously to sunlight.

When the body is wanted for dissection, it should, if possible, be removed from the shell before preservation. If a full dissection is desired, the snail should be killed by drowning in deoxygenated water—which relaxes the animal and expands the tissues. A "squirt" from a soda-water siphon into a glassful of cooled boiled water is a useful source of deoxygenated water. The process may be speeded up by the addition of a few menthol crystals or a few drops of nembutal, or by adding a few drops of alcohol to the water at 10 or 15 minute intervals. Bodies may be preserved in 70% alcohol. For dissection merely to confirm an identification, it is usually safe to extract the body using the technique described for preserving the shell. The pin inserted rarely damages a diagnostic character.

Classification

The classification into families is based on that of Zilch (1959). Species nomenclature is that used in the Conchological Society's census (Kerney 1966), except for a few species recognized in Britain since then. Widely used synonyms are given in the systematic account and in the index.

    Class   GASTROPODA
  Subclass   PROSOBRANCHIA
     Order   MESOGASTROPODA
      Family Pomatiidae
         Pomatias elegans (Müller)

      Family Aciculidae
         Acicula fusca (Montagu)

  Subclass   PULMONATA
     Order   BASOMMATOPHORA
      Family Ellobiidae
         Carychium tridentatum (Risso)
         Carychium minimum (Müller)

     Order   STYLOMMATOPHORA
      Family Succineidae
         Catinella arenaria (Bouchard-Chantereaux)
         Succinea oblonga Draparnaud
         Succinea putris (L.)
         Succinea pfeifferi Rossmässler
         Succinea sarsii Esmark

      Family Cochlicopidae
         Azeca goodalli (Férussac)
         Cochlicopa lubrica (Müller)
         Cochlicopa lubricella (Stabile)

      Family Pyramidulidae
         Pyramidula rupestris (Draparnaud)

      Family Vertiginidae
         Columella edentula (Draparnaud)
         Columella aspera Waldén
         Truncatellina cylindrica (Férussac)
         Truncatellina britannica Pilsbry
         Vertigo pusilla Müller
         Vertigo antivertigo (Draparnaud)
         Vertigo substriata Jeffreys
         Vertigo pygmaea (Draparnaud)

Vertigo geyeri Lindholm
Vertigo moulinsiana (Dupuy)
Vertigo lilljeborgi Westerlund
Vertigo alpestris Alder
Vertigo angustior Jeffreys

Family Pupillidae
Pupilla muscorum (L.)
Lauria cylindracea (da Costa)
Lauria anglica (Férussac)

Family Chondrinidae
Abida secale (Draparnaud)

Family Valloniidae
Acanthinula aculeata (Müller)
Acanthinula lamellata (Jeffreys)
Vallonia costata (Müller)
Vallonia pulchella (Müller)
Vallonia excentrica Sterki

Family Enidae
Ena obscura (Müller)
Ena montana (Draparnaud)

Family Clausiliidae
Marpessa laminata (Montagu)
Clausilia bidentata (Ström)
Clausilia dubia Draparnaud
Clausilia rolphii Leach
Lacinaria biplicata (Montagu)
Balea perversa (L.)

Family Ferussaciidae
Cecilioides acicula (Müller)

Family Endodontidae
Punctum pygmaeum (Draparnaud)
Discus rotundatus (Müller)

Family Zonitidae

Vitrea crystallina (Müller)
Vitrea contracta (Westerlund)
Vitrea diaphana (Studer)
Oxychilus draparnaldi Beck
Oxychilus cellarius (Müller)
Oxychilus alliarius (Miller)
Oxychilus helveticus (Blum)
Retinella radiatula (Alder)
Retinella pura (Alder)
Retinella nitidula (Draparnaud)
Zonitoides excavatus (Alder)
Zonitoides nitidus (Müller)

Family Vitrinidae

Vitrina pellucida (Müller)
Vitrina major Férussac
Vitrina pyrenaica Férussac

Family Euconulidae

Euconulus fulvus (Müller)

Family Bradybaenidae

Fruticicola fruticum (Müller)

Family Helicidae

Helicodonta obvoluta (Müller)
Helicigona lapicida (L.)
Arianta arbustorum (L.)
Theba pisana (Müller)
Helix hortensis Müller
Helix nemoralis L.
Helix aspersa Müller
Helix pomatia L.
Hygromia limbata (Draparnaud)
Hygromia cinctella (Draparnaud)
Hygromia subrufescens (Miller)
Hygromia striolata (Pfeiffer)
Hygromia hispida (L.)
Hygromia subvirescens (Bellamy)
Monacha granulata (Alder)
Monacha cartusiana (Müller)
Monacha cantiana (Montagu)
Helicella caperata (Montagu)
Helicella gigaxii (Pfeiffer)
Helicella virgata (da Costa)
Helicella neglecta (Draparnaud)
Helicella itala (L.)
Helicella elegans (Gmelin)
Cochlicella acuta (Müller)

Key to the British Species

This key allows the identification of all species of land snails living in the wild in the British Isles, including introduced species which have become naturalized, but excluding those restricted to greenhouses or botanic gardens. It is of the usual dichotomous type leading, in the right hand margin, either to the species concerned or to the number of the next relevant couplet. Usually, the determination of characters used requires no more than a good hand lens ($\times 10$), a ruler and a pair of dividers, although a binocular microscope and measuring callipers are better if available.

No key is perfect; this one has been through several editions, and been modified in the light of experience. Nevertheless, there are several important sources of error which cannot be wholly eliminated, and the user should bear the following points in mind.

1. The key is primarily to adult, living specimens. The dimensions, and sometimes other characters, of juveniles are different.
2. Shells which have been empty for some time tend to become opaque and chalky white whatever their previous colouring.
3. Rarely, normally dextrally coiled species produce sinistrally coiled individuals.
4. Species with a closed umbilicus to the shell occasionally produce adult individuals with an open umbilicus. Juveniles of such species often have an open umbilicus (see Fig. 1d).
5. Shell colour is variable. Apart from the regularly polymorphic species, many produce occasional albinos, and a few species with normally white shells may produce brown shelled individuals.
6. Most characters seen on snail-shells are common to many species, and there is a dearth of presence/absence distinctions in this key. Relative measurements are used frequently, and it is important that these be made correctly. We have tried to avoid entirely relative characters, which are useless in the absence of a standard. They do occur, however, and it is hoped that the illustrations will provide the standard.

1. Shell height clearly greater than width 2
   Shell height equal to, or less than width 40
2. Sinistrally coiled 3
   Dextrally coiled 10
3. Adults less than 3 mm high 4
   Adults more than 4 mm high 5
4. Shell thick and striated, with triangular mouth and penultimate whorl wider than last . . . . *Vertigo angustior* (p. 29)
   Shell thin and feebly striated, with squarer mouth and last whorl widest
   Vertigo pusilla (p. 29)
5. Adult more than 14 mm high 6
   Adult less than 14 mm high 8
6. Shell thin, feebly striated and slightly glossy . *Marpessa laminata* (p. 38)
   Shell thick and markedly striated 7
7. Shell pale brown, very heavily striated, about 4 mm wide
   Lacinaria biplicata (p. 38)
   Shell greyish, moderately striated, 3·0–3·5 mm wide   *Clausilia dubia* (p. 38)
8. Width more than 3 mm . . . . *Clausilia rolphii* (p. 38)
   Width less than 3 mm 9
9. Width greatest at level of mouth . . . . . *Balea perversa* (p. 38)
   Width greatest above mouth . . . . . *Clausilia bidentata* (p. 38)
10. Four or more teeth in the mouth* 11
    Less than four or no teeth in the mouth 20
11. Adult height more than 3·0 mm 12
    Adult height less than 3·0 mm 14
12. Adult height less than 4 mm, width about 2 mm . . *Lauria anglica* (p. 32)
    Adult height more than 4 mm, width more than 2 mm 13
13. Shell thick, rough, and strongly striated, about 7 mm high
    Abida secale (p. 32)
    Shell thin and translucent, brown or white, about 5 mm high
    Azeca goodalli (p. 26)
14. Shell markedly and regularly striated, especially on upper whorls
    Vertigo substriata (p. 29)
    Shell only moderately and irregularly striated, if at all 15
15. Usually with 6–9 teeth, with 2 or 3 parietals . *Vertigo antivertigo* (p. 29)
    Usually less than 6 teeth; only 1 parietal, or occasionally with a very small second one 16

---

* *Vertigo geyeri* is variable, and specimens with anything between 0 and 4 teeth may occur. Specimens with 0–3 teeth will key out under couplet **22**, leading to *Cecilioides* and *Carychium*, from which its brown shell and general shape distinguish it.

16. Large (about 2·5 mm high). Height of body whorl more than 2× that of the spire. Massive flaring lip . . . . . . *Vertigo moulinsiana* (p. 30)
    Smaller (about 2·0 mm high). Height of body whorl less than 2× that of the spire. Lip not massive or much flared 17
17. Globular, breadth about ⅔× height . . . *Vertigo lilljeborgi* (p. 30)
    More cylindrical or pointed, breadth about ½× height* 18
18. Usually 5 (sometimes 4) teeth, very weakly striated and relatively thick lip . . . . *Vertigo pygmaea* (p. 29)
    Four (or less) teeth, moderately striated, with thinner lip† 19
19. Very cylindrical, yellowy-brown shell with 4 teeth . *Vertigo alpestris* (p. 30)
    More conical, darker brown shell with 0–4 teeth† . . *Vertigo geyeri* (p. 29)
20. Height 4 mm or less, or if more, then less than 1·5 mm wide . . . 21
    Height 5 mm or more, or if less, then more than 2·0 mm wide . . . 30
21. Height of last whorl more than ½× shell height 22
    Height of last whorl less than ½× shell height 25
22. Height more than 3 mm 23
    Height less than 3 mm 24
23. Shell 4–5 mm high, about 1·3 mm wide, no lip or teeth
                                           Cecilioides acicula (p. 37)
    Shell 3·5–4 mm high, with thickened lip and one tooth
                                           Lauria cylindracea (p. 32)
24. Height usually less than 2× width, shallow sutures, internal columellar folds smooth . . . . *Carychium minimum* (p. 23)
    Height usually more than 2× width, deeper sutures, internal columellar folds irregular . . . . *Carychium tridentatum* (p. 23)
25. Height 3 mm or less, width 1 mm or less 26
    Larger in one or both dimensions 28
26. Shell glossy and translucent with few well marked vertical striations and no lip . . . . *Acicula fusca* (p. 22)
    Shell with many fine striations and lip on mouth 27
27. Lip well developed. 2 or 3 small teeth in mouth set back from lip
                                           Truncatellina britannica (p. 28)
    Lip weakly developed, no teeth . . . . *Truncatellina cylindrica* (p. 28)
28. With a thickened white rib behind mouth . . *Pupilla muscorum* (p. 32)
    Without such a rib 29

---

* *V. lilljeborgi* is very variable, and specimens may occasionally resemble other species rather closely.
† *Vertigo geyeri* is variable, and specimens with anything between 0 and 4 teeth may occur. Specimens with 0–3 teeth will key out under couplet 22, leading to *Cecilioides* and *Carychium*, from which its brown shell and general shape distinguish it.

29. Height up to 3 mm, cylindrical, with very faint striations
    . . . . *Columella edentula* (p. 28)
    Height about 2·5 mm, more tapering, and with more marked striations, especially on the upper whorls . . . . . *Columella aspera* (p. 28)
30. Shell massive and opaque with heavy horizontal striations and an operculum . . . . *Pomatias elegans* (p. 22)
    Shell without operculum or heavy horizontal striations 31
31. Shell uniform opaque brown with whitish lip 32
    Shell not opaque brown, with or without lip 33
32. Shell more than 12 mm high . . . . *Ena montana* (p. 36)
    Shell less than 12 mm high . . . . *Ena obscura* (p. 36)
33. Height of mouth nearly or more than ½ × height of shell 34
    Height of mouth ⅓ or less shell height 38
34. Height of shell mouth approx. ½ × shell height 35
    Height of shell mouth much more than ½ × shell height 36
35. Columellar face of mouth slightly angled, adult not more than about 6 mm high. Vas deferens does not thicken into an epiphallus before joining penis . . . . *Catinella arenaria* (p. 24)
    Columellar face of mouth smoothly curving, adult up to about 10 mm high. Vas deferens thickens into epiphallus before joining penis
    . . . . *Succinia oblonga* (p. 24)
36. By dissection: jaw with three cusps . . . . . *Succinia putris* (p. 24)
    jaw with one central cusp only* 37
37. Vagina short and straight . . . . . . *Succinia pfeifferi* (p. 24)
    Vagina long and bent into S shape . . . . . *Succinia sarsii* (p. 24)
38. Shell white, with or without dark blotches or bands which may nearly obliterate the white, no lip . . . . . *Cochlicella acuta* (p. 60)
    Shell pale brown and transparent, adults with lip 39
39. Sutures comparatively deep. Height usually more than 5·5 mm, wide in proportion to height . . . . *Cochlicopa lubrica* (p. 26)
    Sutures comparatively shallow. Height usually less than 6·0 mm, narrow in proportion to height† . . . . . . *Cochlicopa lubricella* (p. 26)

---

* Distinctions can sometimes be made between *S. putris* and *S. pfeifferi* and *S. sarsii* on shell characters; *S. putris* has the largest body whorl in proportion to the spire, and the most shallow sutures. Typical specimens are shown in Fig. 4 but the distinctions are not wholly reliable.

† The distinctions between *C. lubrica* and *C. lubricella* are the most difficult to make. There seem to be no reliable internal differences, and individuals with intermediate shell characters are found quite often.

40.	Shell with marked keel (Fig. 1c)	41
	Shell with slight keel or none*	43
41.	Width about 15 mm or more, width 2× height . *Helicigona lapicida* (p. 50)	
	Width less than 15 mm, and less than 2× height	42
42.	Shell cone-shaped with very flattened base . . *Helicella elegans* (p. 60)	
	Shell more rounded, with very small umbilicus . *Hygromia cinctella* (p. 54)	
43.	Shell hairy (hairs may be abraded, but usually remain in the umbilicus)	44
	Shell not hairy (but may have spines)†	47
44.	Apex below level of last whorl . . . . *Helicodonta obvoluta* (p. 50)	
	Apex above level of last whorl	45
45.	Umbilicus minute	46
	Umbilicus ¼–⅓ width of last whorl opposite mouth, hairs curved‡	
	Hygromia hispida (p. 54)	
46.	Shell greenish with sparse hairs, whorls expand very rapidly	
	Hygromia subvirescens (p. 54)	
	Whorls do not expand rapidly, dense covering of straight hairs	
	Monacha granulata (p. 56)	
47.	Adult shell with umbilicus completely closed	48
	Adult shell with open umbilicus, even if very small§	54

---

* Juveniles of *Theba pisana*, *Hygromia striolata* and *Discus rotundatus* may appear strongly keeled.
† Juveniles of *Hygromia striolata*, *Monacha cantiana* and *M. cartusiana* are hairy for some time after hatching.
‡ Some varieties of *H. hispida*, especially common in the Midlands, may have a rather small umbilicus. The shape of the hairs is diagnostic.
§ Juvenile *Helix* spp have an open umbilicus, but they can be distinguished from adults by the lack of a lip. In some populations a small proportion of adults may be umbilicate.

48. Adult with strongly developed lip, always more than 12 mm wide . . **49**
    Adult without lip, less than 7 mm wide **51**
49. Shell usually dull brown or yellow in colour, usually more than 30 mm wide. Bands, if present, usually irregular and flecked . . *Helix aspersa* (p. 52)
    Shell brightly coloured, yellow, pink or brown, with up to five bands or none, usually less than 25 mm wide **50**
50. Lip usually white. Dart-flanges bifurcated, usually 4 or more branches to each mucous gland . . . . *Helix hortensis* (p. 52)
    Lip usually dark brown. Dart-flanges simple, usually 3 or less branches to each mucous gland* . . . . *Helix nemoralis* (p. 52)
51. Shell tightly coiled, brown, about 3·5 mm wide . *Euconulus fulvus* (p. 49)
    Shell with rapidly expanding whorls, transparent, colourless or slightly greenish **52**
52. Last whorl grossly expanded, mouth more than ¾ shell width
    Vitrina pyrenaica (p. 48)
    Last whorl less expanded, mouth about ⅔ shell width **53**
53. Adults present in winter, body grey with black head, oviduct simple, spermathecal duct joins oviduct near junction with penis
    Vitrina pellucida (p. 48)
    Adults present in summer, body all black, oviduct with characteristic swelling, at which point the spermathecal duct joins it
    Vitrina major (p. 48)
54. Adult shell less than 5 mm wide **55**
    Adult shell more than 5 mm wide **66**
55. Adult shell with large lip round mouth **56**
    Adult shell without lip **58**
56. Shell with marked radial ribs . . . . . . *Vallonia costata* (p. 34)
    Shell without such ribs **57**
57. Lip seen from outside and behind only slightly flared and umbilicus markedly eccentric. Shell smooth . . . . . *Vallonia excentrica* (p. 34)
    Lip seen from outside and behind markedly flared, umbilicus slightly eccentric and shell slightly striated . . . *Vallonia pulchella* (p. 34)
58. Adult shell width ≃ to height **59**
    Adult shell width clearly greater than height **60**
59. Shell with casing of sculptured spines . . . *Acanthinula aculeata* (p. 34)
    Shell without such spines . . . . . . *Acanthinula lamellata* (p. 34)

---

* Lip colour is usually reliable. White lipped *H. nemoralis* occur rarely in many populations, most often in Irish and Pennine populations. Dark-lipped *H. hortensis* occur sporadically, mainly in populations from S.W. England.

60. Shell tightly coiled, smooth and glossy, with umbilicus ¼ or less width of last whorl opposite mouth **61**
Umbilicus more than ¼ width of last whorl opposite mouth; if shell glossy, then not tightly coiled **63**
61. Umbilicus minute . . . . *Vitrea diaphana* (p. 42)
Umbilicus about ¼ width of last whorl opposite mouth **62**
62. Relatively flattened spire, last whorl about $1\frac{1}{3} \times$ width of the next, umbilicus circular, and underside of shell and mouth edge flattened. Maximum with about 2·5 mm . . . . *Vitrea contracta* (p. 41)
Relatively pointed spire, last whorl about $1\frac{1}{2} \times$ width of the next, umbilicus eccentric, underside of shell rounded. Maximum width about 3·3 mm
Vitrea crystallina (p. 41)
63. Shell with regular striations, usually brown in colour, glossy or rough and opaque **64**
Shell smooth, with feeble and irregular striations, usually white*
Retinella pura (p. 46)
64. Shell very glossy, transparent and brown, with rapidly expanding whorls and fine very regular radial striations . . *Retinella radiatula* (p. 46)
Shell not glossy, and more tightly coiled **65**
65. Shell about 3 mm wide, with heavy striations and deep sutures
Pyramidula rupestris (p. 27)
Shell about 1·5 mm wide, with finer striations and shallower sutures
Punctum pygmaeum (p. 40)
66. Shell 35 mm or more wide . . . . . . *Helix pomatia* (p. 52)
Shell less than 25 mm wide **67**
67. Adult shell 15 mm or more wide, with well defined white lip nearly closing umbilicus, usually brown, and nearly always flecked. Often with dark central band . . . . *Arianta arbustorum* (p. 50)
Shell without lip (a rib may be found behind mouth edge), not flecked **68**
68. Adult shell less than 14 mm wide, thin, translucent or transparent, smooth, glossy or waxy, with striations feeble if present. No thickened ribs behind mouth, no bands or blotches **69**
Adult size variable, shells opaque, chalky white or brown, never glossy. Bands, blotches and marked striations may be present, as may a rib behind mouth **76**
69. Umbilicus minute, shell pale brown, about 9 mm wide, width about 1·6× height . . . . *Hygromia subrufescens* (p. 54)
Umbilicus at least ¼ width of last whorl opposite mouth, width about 1·8–2·0× height **70**

---

* *Retinella pura* is sometimes brown, in which case confusion with juvenile *R. nitidula* is possible. By comparison, *R. pura* is more tightly coiled, and has finer striations.

70. Umbilicus ½ width of last whorl opposite mouth; waxy, or if glossy then with relatively deep sutures **71**
Umbilicus less than ½ width of last whorl opposite mouth, glossy with shallow sutures **73**

71. Whorls expand rapidly, shell waxy brown . . . *Retinella nitidula* (p. 46)
Whorls expand slowly, shell glossy, but may be heavily pitted and eroded **72**

72. Umbilicus about ¼ total width of shell and very open, shell relatively more convex above. Mantle covered with white spots
Zonitoides excavatus (p. 45)
Umbilicus about ⅕ total width of shell, less cavernous. Shell flatter above. Mantle with black spots . . . . . . *Zonitoides nitidus* (p. 45)

73. Shell 12 mm or more wide, last whorl expands rapidly near mouth. Shell usually less glossy and more striated than other *Oxychilus* species
Oxychilus draparnaldii (p. 43)
Shell less than 11 mm wide, very glossy, with very feeble striations if any **74**

74. Adult shell less than 7 mm wide, umbilicus proportionately large. Animal smells strongly of garlic when irritated . . . *Oxychilus alliarius* (p. 43)
Adult shell more than 7 mm wide, umbilicus proportionately smaller; never smells strongly of garlic **75**

75. Shell very pale brown, umbilicus relatively large, mantle pale grey. Shell usually very flattened . . . . *Oxychilus cellarius* (p. 44)
Shell darker brown, usually more convex above, and with relatively narrow umbilicus. Mantle black . . . . . . *Oxychilus helveticus* (p. 44)

76. Shell markedly and regularly striated **77**
Shell weakly striated or not at all **79**

77. Shell very flattened (width about 2× height) with umbilicus about ⅔ width of last whorl opposite mouth. Shell usually brown with pattern of dark blotches, but may be completely white . . *Discus rotundatus* (p. 40)
Shell not so flattened, umbilicus less than ½ width of last whorl opposite mouth. Shell usually with a chalky white background covered with dark bands or blotches **78**

78. Umbilicus relatively small, somewhat conical shell with coarse striations (⅔–¾ as many per unit distance as next species) *Helicella caperata* (p. 58)
Umbilicus relatively large, flatter shell, finer striations
Helicella gigaxii (p. 58)

79. Umbilicus minute (less than 1/10 width of last whorl opposite mouth), shell rarely more than 13 mm wide **80**
Umbilicus at least ⅛ width of last whorl opposite mouth, shell often larger than 13 mm wide **81**

80. Shell width less than $1\frac{1}{2}\times$ height, usually brown, and often with one central dark or light band. No lip, but white rib behind mouth edge
Hygromia limbata (p. 54)
Shell width usually more than $1\frac{1}{2}\times$ height, whitish, with dark colouration round mouth-edge . . . . *Monacha cartusiana* (p. 56)
81. Umbilicus $\frac{1}{3}$ or more width of last whorl opposite mouth 82
Umbilicus $\frac{1}{4}$ or less width of last whorl opposite mouth 85
82. Umbilicus about $\frac{1}{3}$ width of last whorl opposite mouth, never with dark bands and blotches 83
Umbilicus $\frac{1}{2}$ or more width of last whorl opposite mouth, usually with dark bands or blotches . . . . 84 *Helicella*
83. Shell rather conical above, and with slight keel, brown or yellowish
Hygromia striolata (p. 54)
Shell very rounded, without keel, yellowish white
Fruticicola fruticum (p. 49)
84. Very flattened shell (width $1\frac{3}{4}$ or more × height) rib small if present
Helicella itala (p. 58)
Shell not so flattened (width about $1\frac{2}{3}\times$ height) rib inside mouth well developed . . . . *Helicella neglecta* (p. 58)
85. Shell width $1-1\frac{1}{4}\times$ height; bands and blotches usual 86
Shell width $1\frac{1}{3}-1\frac{2}{3}\times$ height, never with dark bands, but rufous flush may extend backwards from mouth . . . . *Monacha cantiana* (p. 56)
86. Shell with pinkish flush inside mouth, width up to 20 mm
Theba pisana (p. 50)
Shell mouth may be brown or ginger, but not pink . *Helicella virgata* (p. 58)

Systematic Part

(*Note in this part all measurements are only approximate*)

Sub Class: Prosobranchia
Order: Mesogastropoda

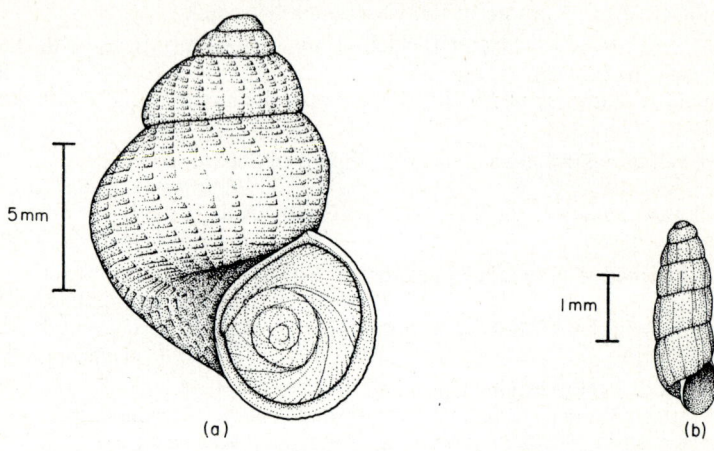

FIG. 3. (a) Pomatiidae: *Pomatias elegans*; (b) Aciculidae: *Acicula fusca*.

Family POMATIIDAE

Pomatias elegans (Müller) (Fig.3a) K30

Shell very thick and opaque with strong spiral striations; faintly purple, with two or three faint, darker spiral bands often present. Sutures deep. Operculum with spiral growth lines. Sexes separate, the female being slightly larger than the male. Height 12–16 mm, width 8–12 mm. In a variety of habitats, usually on loose soil, and always on calcareous soil. Southern and central England, just extending to N. Wales.

Family ACICULIDAE

Acicula fusca (Montagu) (Fig. 3b) K26

(*Acme fusca* (Montagu))

Shell narrow and nearly cylindrical, with few, widely spaced vertical striations. Usually pale brown, but may be white or yellowish. Operculum thin and horny. Height 2·0–2·5 mm, width about 0·5 mm. Usually a woodland species, found locally throughout the British Isles.

Sub Class: *Pulmonata*
Order: *Basommatophora*
Family ELLOBIIDAE

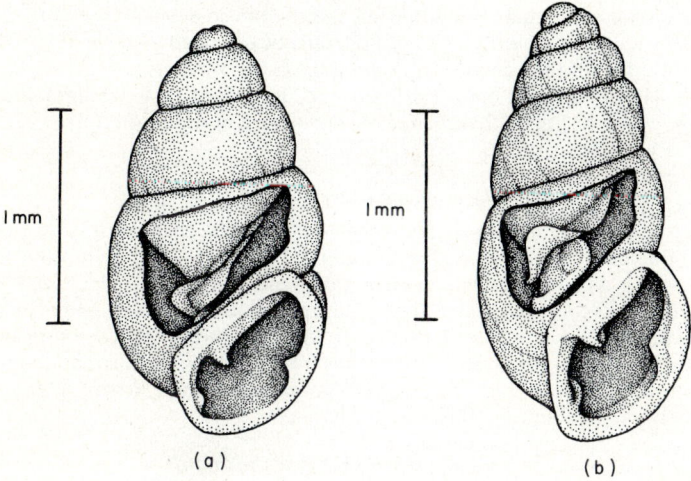

FIG. 4. Ellobiidae: (a) *Carychium minimum*, (b) *C. tridentatum*, both cracked open to show the internal folds on the columella.

Carychium minimum (Müller) (Fig. 4*a*)

Shell similar to that of *C. tridentatum*, more rounded, with height usually less than twice width, and with striations very faint and irregular. Continuation of parietal tooth inside body whorl circles the columella in a smooth curve. Height 1·8–2·0 mm, width about 1 mm. Found usually in damp habitats—fens, carrs and damp woods—but may often be found with its congener. Widespread in the British Isles.

Carychium tridentatum (Risso) (Fig. 4*b*)

Shell white and steeply conical, with deeper sutures than *C. minimum*, and with fine striations. Mouth with three teeth and a thickened lip. The continuation of the parietal tooth inside the body whorl circles round the columella in an irregular and angular way. Height of shell usually more than twice the width. Height about 2 mm, width 1 mm. Found in a wide range of habitats, especially on calcareous soils, throughout the British Isles.

Order: Stylommatophora
Family SUCCINEIDAE
Catinella arenaria (Bouchard-Chantereaux) (Fig. 5a)

Thin, pale brown transparent shell, with about three rapidly expanding whorls and deep sutures. Unlike the following species, the columellar face of the mouth is weakly angled. Certainty of diagnosis must be obtained by dissection—the vas deferens does not thicken prior to joining the penis, which is not surrounded by a sheath. Height up to 6 mm, width 3·5 mm. Restricted to wet habitats, usually close to water. In England, recorded only at present at Braunton Burrows, Devon, but more widespread in S. and W. Ireland.

Succinea oblonga Draparnaud (Fig. 5b)

The shell is very similar in shape to the above species, thicker and more opaque than other succineids, and with more marked striations. The columellar face of the mouth is a continuous curve. Height up to 10 mm, width to 5·5 mm. By comparison with the above species, the vas deferens thickens into an epiphallus prior to joining the penis, which is itself enclosed in a sheath. Often found in drier situations than its congeners, but mostly restricted to the western counties of Britain, usually near the coast, but commoner in Ireland, where it may occur commonly inland and in relatively dry habitats.

Succinea putris (L.) (Fig. 5c and d)

Shell glossy, thin and translucent, with very rapidly expanding whorls. The spire is very small, and the sutures shallower than in the following species. The size of the spire, and roundness of the mouth may serve to distinguish this species from the two following, but certain diagnosis may be made by examination of the jaw, which has three distinct teeth. Height usually 17–20 mm, width 10–11 mm, but occasionally larger. Restricted to wet habitats—watermeadows, fens, river-banks, reed and sedge beds. Found throughout the British Isles.

Succinea pfeifferi Rossmässler (Fig. 5e and f)

Shell very similar to *S. putris* but usually with a larger spire, and a mouth taller in proportion to its breadth. The jaw has a single median tooth. Although the shell is broader in proportion to height, and less heavily striated than that of *S. sarsii*, reliable determination must depend on the genitalia. The vagina is short and straight. Height 10 mm, width 6 mm. Habitats similar to those of *S. putris*; found throughout the British Isles.

Succinea sarsii Esmark (Fig. 5g)
(*S. elegans* Risso)

Shell very similar to those of the two preceeding species, but it is more heavily striated, and slightly higher in proportion to its width. Height 14–15 mm, width 6–7 mm. Reliably distinguished from *S. putris* by possessing but one median tooth in the jaw, and from *S. pfeifferi* by having a long, thickened vagina bent back over itself *in situ*. Restricted to marshy habitats in S.E. England.

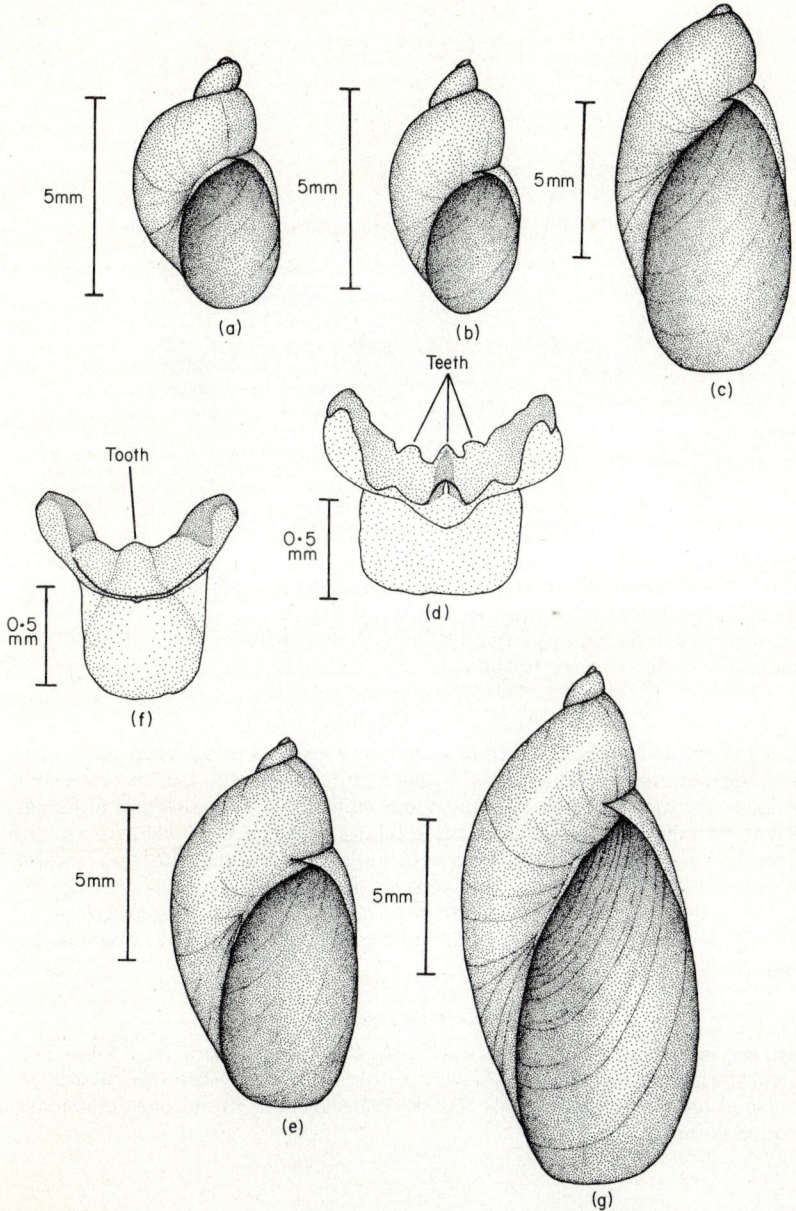

FIG. 5. Succineidae: (a) *Catinella arenaria*; (b) *Succinea oblonga*; (c) *S. putris* shell; (d) *S. putris* jaw; (e) *S. pfeifferi* shell; (f) *S. pfeifferi* jaw; (g) *S. sarsii*.

BRITISH LAND SNAILS
Family COCHLICOPIDAE

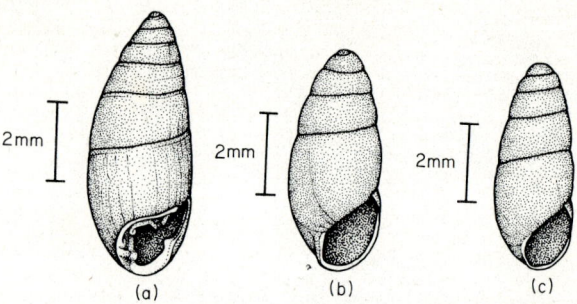

FIG. 6. Cochlicopidae: (a) *Azeca goodalli*, (b) *Cochlicopa lubrica*, (c) *C. lubricella*.

Azeca goodalli (Férussac) (Fig. 6a)

Shell smooth and glossy, with very shallow sutures, and with from 6–9 small teeth in the mouth. Height 6–7 mm, width about 2·5 mm. Usually pale brown, but albino shells are quite frequent. A widely distributed species but rare and local, occurring in woods and other damp habitats in most parts of England and Wales and in a few sites in central Scotland.

Cochlicopa lubrica (Müller) (Fig. 6b)

Shell glossy and transparent, yellowish to pale brown, subcylindrical, and with a lip at the mouth edge. Height 5·0–7·5 mm, width 2·4–3·0 mm. This species is very similar to the following one, and anomalous individuals are often found, although they are usually only a small proportion of any one sample. *C. lubrica* is usually larger, has more tumid whorls and deeper sutures, is less cylindrical and is wider in proportion to its height than *C. lubricella*. The distal end of the penial appendage is longer and more swollen than in *C. lubricella*. Found in a wide range of habitats throughout the British Isles, but usually replaced by *C. lubricella* in the driest places.

Cochlicopa lubricella (Stabile) (Fig. 6c)

Shell very similar to that of *C. lubrica*. Height 4·5–7·0 mm, width 2·0–2·5 mm. For diagnostic characters see above. Widely distributed in the British Isles, usually in drier and more exposed habitats than its congener, but mixed populations are frequently found.

Family PYRAMIDULIDAE

 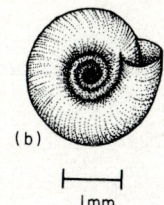 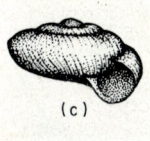

FIG. 7. Pyramidulidae: *Pyramidula rupestris*, top (a), bottom (b) and (c) side views.

Pyramidula rupestris (Draparnaud) (Fig. 7a, b and c)

Opaque, somewhat globular shell, whorls tightly coiled, with deep sutures, a large umbilicus and heavy striations. Colour dark brown to purple. Height 1·5–1·8 mm, width 2–3 mm. Characteristic of rocks—found on cliffs, screes, boulders and walls, usually on limestone. Widely distributed in the British Isles.

Family VERTIGINIDAE

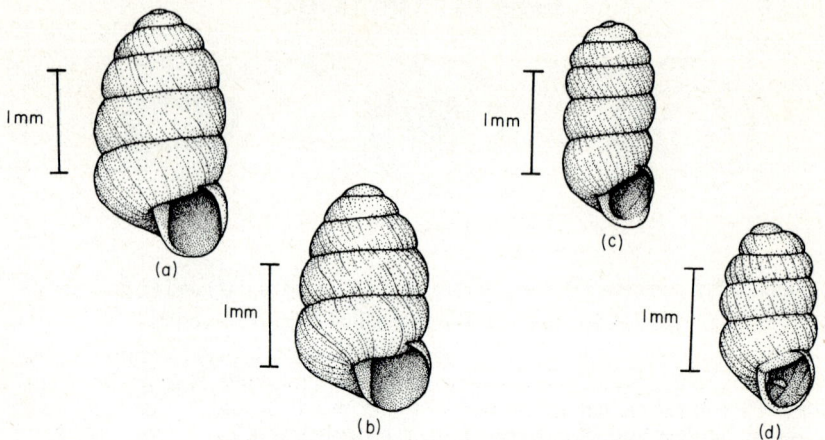

FIG. 8. Vertiginidae, genera *Columella* and *Truncatellina*: (a) *Columella edentula*, (b) *C. aspera*, (c) *Truncatellina cylindrica*, (d) *T. britannica*.

Columella edentula (Draparnaud) (Fig. 8a)

Shell cylindrical, brown, with deep sutures and faint irregular striations, no lip or teeth. Height up to 3 mm, width 1·2–1·5 mm. Distinguished from *C. aspera* by more cylindrical shape and less heavy striations on upper whorls. Habitats very varied, but found most frequently in damp woods and in wetlands. Found throughout the British Isles.

Columella aspera Waldén (Fig. 8b)

Very similar to the previous species, but with thicker, more tapering shell, more heavily striated, especially on upper whorls. Height up to 2·5 mm, width about 1·4 mm. As this species has only recently been recognized from Britain, its range and habitats are as yet uncertain; probably more northern in distribution than *C. edentula*.

Truncatellina cylindrica (Férussac) (Fig. 8c)

Tiny cylindrical shell with fine regular striations. Rounded mouth with a poorly developed lip. Height about 1·8 mm, width about 0·8 mm. Distinguished from *T. britannica* by the absence of teeth, the finer striations, and the more feebly developed lip. Habitat: dry grassland or exposed ground; found in a few scattered localities from central Scotland to S. England, mostly on the east side of the country.

Truncatellina britannica (Pilsbry) (Fig. 8d)

Very similar to *T. cylindrica*, but with a better developed lip, somewhat turned-out, and usually 3 small teeth set well back in the mouth, of which the palatal is usually the most noticeable in front view. More coarsely striated than *T. cylindrica*. Height 1·7–2·0 mm, width 0·8–0·9 mm. Habitats similar to those of *T. cylindrica*, but restricted to coastal areas in S. England; rare, localized.

Vertigo pusilla (Müller) (Fig. 9a)

Sinistral, yellowy-brown shell, feebly striated, with a somewhat square mouth with 6 or 7 teeth (2 parietal, 2 columellar and 2 palatal, with a small extra tooth sometimes present between the palatals and columellars). Height about 2·0 mm, width about 1·0 mm. Widely distributed in the British Isles, but rare and local, usually on walls or other dry but shaded places.

Vertigo angustior Jeffreys (Fig. 9b)

Sinistral, yellowish-brown shell, glossy with strong oblique striations. Mouth nearly triangular with 4–5 teeth (2 parietal, 1 columellar, 1–2 palatal) and with a well developed lip. Height 1·5–2·0 mm, width 0·8–1·0 mm. Found in wetlands in Norfolk and more widespread in Ireland. Other localities in Britain seem to yield only subfossil shells.

Vertigo antivertigo (Draparnaud) (Fig. 9c)

Dextral dark brown shell, slightly glossy. Mouth with well developed lip and at least 6 teeth (2 parietal, 2 columellar and 2 palatal) with 2–3 extra teeth often present. Height 2·0–2·2 mm, width 1·2–1·3 mm. In wetlands and on river banks, especially on sedges and rushes. Widely distributed in the British Isles.

Vertigo substriata Jeffreys (Fig. 9d)

Dextral, thin, pale brown shell with marked regular striations most clearly shown on upper whorls. Mouth with thin lip and usually 5–6 teeth (2 parietal, 2 columellar and 2 palatal) but 1 or 2 teeth may be missing. Height 1·7–2·0 mm, width about 1·0 mm. In wetlands, usually near ground level, but also in woods, especially in Ireland. Widespread in the British Isles, but commoner in the north and in Ireland.

Vertigo pygmaea (Draparnaud) (Fig. 9e)

Dextral brown shell, very feebly striated, if at all. Mouth with thickened lip and 4–5 teeth (1 parietal, 1 columellar and 2–3 palatals), which are thick and white. Height 1·9–2·2 mm, width 0·9–1·2 mm. Widespread in open grassland habitats, especially on calcareous soil, but also in some wetlands. Throughout the British Isles.

Vertigo geyeri Lindholm (Fig. 9f and g)
(*V. genesii* Gredler)

Dextral, brown shell, tapering towards the spire, moderately striated. Mouth with weakly developed lip and 0–4 thin brown teeth (1 parietal, 1 columellar, 2 palatal)—specimens with no teeth quite frequent. Shell thinner, more striated and with less thickening round the mouth than in *V. pygmaea* and less globular than *V. lilljeborgi*. Height 1·6–1·8 mm, width about 1·0 mm. Restricted to damp habitats in central Ireland, very localized.

Vertigo moulinsiana (Dupuy) (Fig. 9h)

The largest *Vertigo*, about 2·3–2·6 mm high, 1·3–1·5 mm wide. Dextral, yellowish-brown, with body whorl twice the height of the spire, and with a massive turned out lip. 4–5 teeth (1 parietal, 1 columellar, 2–3 palatal). Found in wetlands, especially on reeds and sedges, in the southern half of England and in some Irish counties.

Vertigo lilljeborgi Westerlund (Fig. 9i)

Dextral; somewhat globular shell with large, wide body whorl, mid-brown, and with a much less developed lip than in *V. moulinsiana*. 4–5 teeth (1 parietal (with occasionally a second very small one), 1–2 columellar and 2 palatal). Height 2·0–2·2 mm, width 1·2–1·4 mm. Found on rotting vegetation and stones round the margins of lakes and rivers in Ireland, Scotland, the English Lake District and N. Wales.

Vertigo alpestris Alder (Fig. 9j)

Dextral, very cylindrical, pale yellow-brown shell with moderate striations. 4 teeth (1 parietal, 1 columellar, 2 palatal). Height about 2 mm, width about 1 mm. Found on rocks and walls in N. England, N. Wales and Mull. Localized.

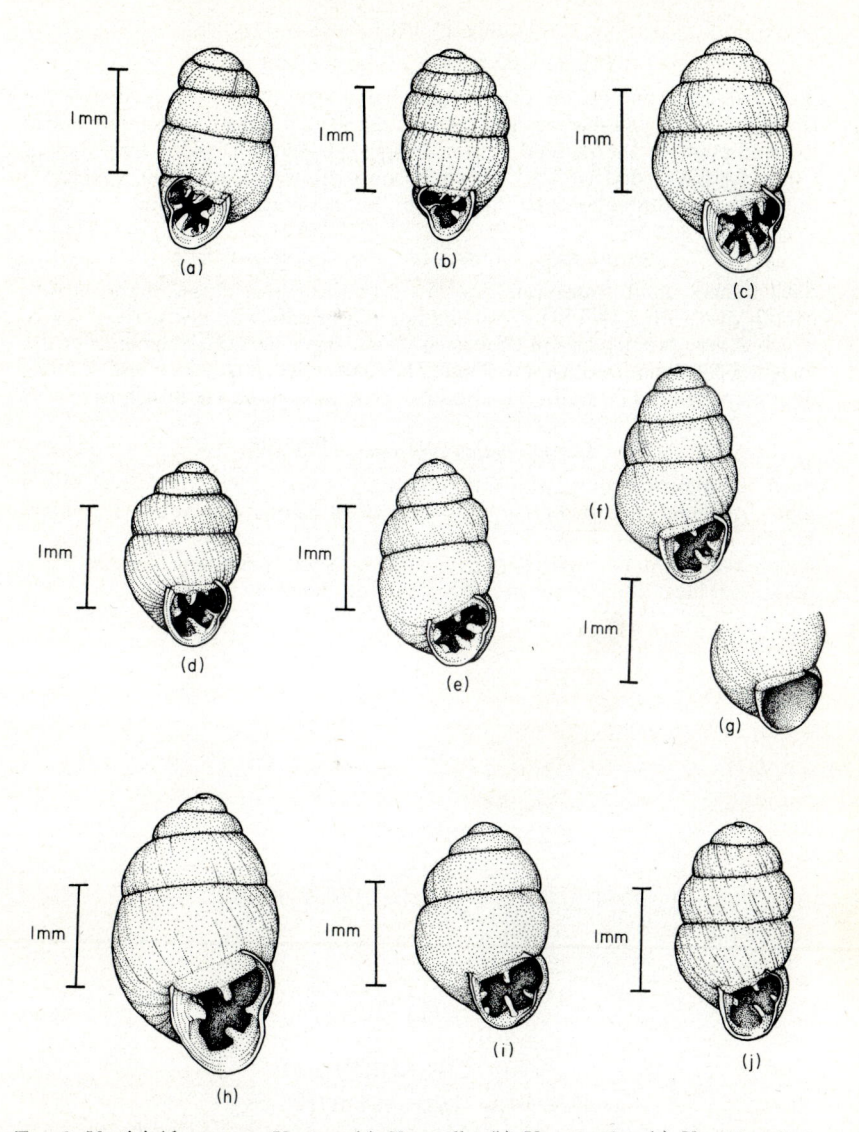

FIG. 9. Vertiginidae, genus *Vertigo*: (a) *V. pusilla*, (b) *V. angustior*, (c) *V. antivertigo*, (d) *V. substriata*, (e) *V. pygmaea*, (f) and (g) *V. geyeri* shells with and without teeth, (h) *V. moulinsiana*, (i) *V. lilljeborgi*, (j) *V. alpestris*.

Family PUPILLIDAE
Pupilla muscorum (L) (Fig. 10a and b)

Shell cylindrical and brown, with a thickened lip backed by a conspicuous white rib behind, clearly visible on the outside of the shell. 0–3 teeth, commonly one in the mid-point of the parietal region. Height 3·0–3·6 mm, width 1·6–1·9 mm. Characteristic of dry habitats, especially short calcareous grasslands and sand-dunes. Occurs throughout the British Isles, but is rarer in the north.

Lauria cylindracea (da Costa) (Fig. 10c)

Shell thinner, more translucent and less cylindrical than *P. muscorum*, brown. Mouth oval with a well marked white lip, and usually with one parietal tooth, which is very prominent and connected to the upper part of the columellar lip. Height 3·5–4·3 mm, width about 2 mm. Habitats varied, but often found in rocky or stoney sites and on walls, but also in woods, throughout the British Isles.

Lauria anglica (Férussac) (Fig. 10d)

Shell rather oval, brown, with shallower sutures than in *L. cylindracea*, with a more triangular mouth containing 4 teeth (2 parietal, 1 columellar, 1 palatal). Height 3·5–4·0 mm, width about 2 mm. Habitat—generally in wet places—marshes and woods especially in Scotland, N. England and Ireland, with occasional local populations in S. England and Wales.

Family CHONDRINIDAE
Abida secale (Draparnaud) (Fig. 10e)

Shell a long cylinder, but with a pointed spire, and brown. Usually 7–8 teeth (2–3 parietals, 2 columellars, 4 palatals). Height 7–9 mm, width about 3 mm. Usually restricted to rocky hillsides, occasionally in non-rocky woods, always on limestone or chalk. Central S. England and the Lake District and N.W. Pennines only.

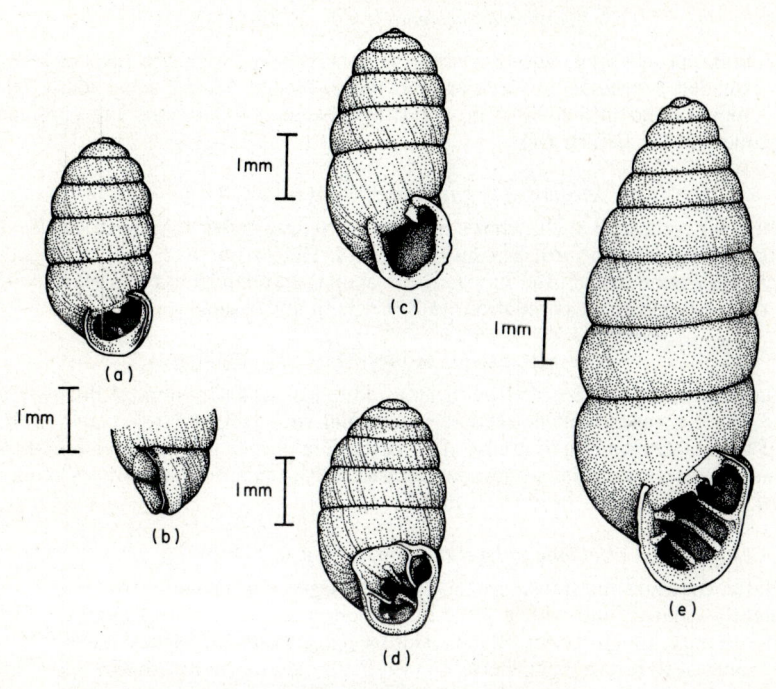

FIG. 10. (a), (b), (c) and (d) Pupillidae: *Pupilla muscorum* (a) front view of whole shell, (b) side view of last whorl showing white thickened rib behind mouth, (c) *Lauria cylindracea*, (d) *L. anglica*; (e) Chondrinidae: *Abide secale*.

Family VALLONIIDAE
Acanthinula aculeata (Müller) (Fig. 11a)

Globular, brown shell, with thin white lip. Whorls bear strong ribs, each of which is extended outwards into a triangular spine. Height 1·8–2·0 mm, width 2·0–2·3 mm. Found in leaf litter in woods and hedges. Widespread and common throughout the British Isles.

Acanthinula lamellata (Jeffreys) (Fig. 11b)

Globular, yellowish shell, whorls much more tightly coiled than in *A. aculeata*. Surface regularly ribbed, but lacking spines. Height about 2 mm, width 2·0–2·3 mm. Found in leaf litter in woodland. Scotland, Ireland and N. England, with a few scattered relict populations in S. England and Wales.

Vallonia costata (Müller) (Fig. 11c, d and e)

Flattened white to grey shell with large umbilicus and pronounced thick white lip. Shell covered with marked regular radial ribs. Height 1·2–1·3 mm, width 2·5 mm. Characteristic of dry habitats, calcareous grassland and dunes and at the base of walls and under stones, but occasionally found elsewhere. Widespread in the British Isles, but rarer in Scotland.

Vallonia pulchella (Müller) (Fig. 11f, g and h)

Flattened, white and translucent shell with a large, somewhat eccentric umbilicus, slightly striated, and with a large lip, which is conspicuously expanded at the mouth-edge. Height about 1·2 mm, width 2·4 mm. Found in rather damp habitats, in contrast to its two congeners. Widely distributed in the British Isles.

Vallonia excentrica Sterki (Fig. 11i, j and k)

Flattened, transparent shell, umbilicus very eccentric, glossy, without striations, and with a large white lip expanded only very slightly at the mouth edge. Height about 1·1 mm, width 2·2–2·3 mm. Like *V. costata*, characteristic of dry, grassy, habitats, throughout the British Isles.

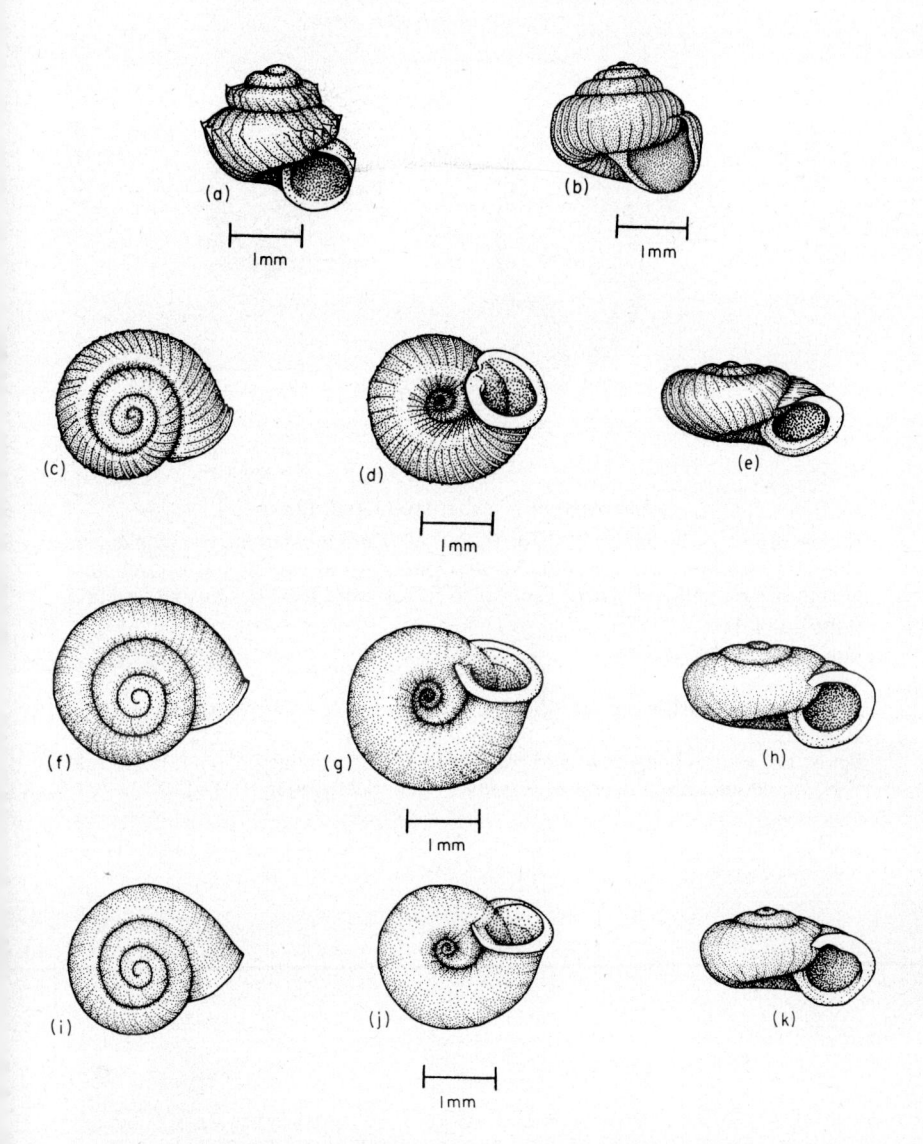

FIG. 11. Valloniidae: (a) *Acanthinula aculeata*; (b) *A. lamellata*; (c), (d) and (e) top bottom and side views of *Vallonia costata*; (f), (g) and (h) top, bottom and side views of *V. pulchella*; (i), (j) and (k) top, bottom and side views of *V. excentrica*.

Family ENIDAE

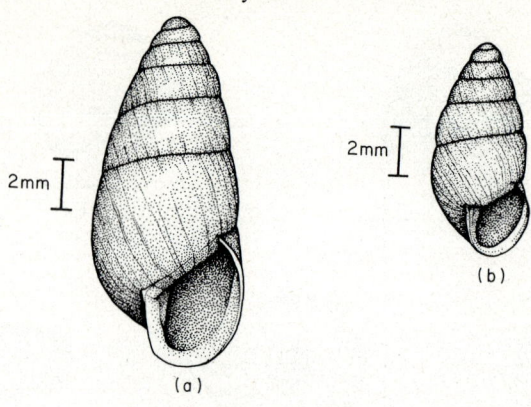

FIG. 12. Enidae: (a) *Ena montana*, (b) *E. obscura*.

Ena montana (Draparnaud) (Fig. 12*a*)

Richer brown shell, similar in shape to that of *E. obscura* but larger and glossier. Height 13–16 mm, width 6 mm. Mostly found in old woods, but occasionally also in hedges. A local species, known from scattered localities in southern and central England.

Ena obscura (Müller) (Fig. 12*b*)

Brown conical shell with white lip and no teeth; shell smaller and rougher than *E. montana*. Height 7–9 mm, width, 3·5–4·0 mm. On trees and in litter in woodlands, but also in hedgerows and rocky places. Throughout England, Wales and S. Scotland, but absent from the extreme North. Scattered in Ireland.

Family FERUSSACIIDAE

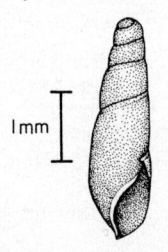

FIG. 13. Ferussacidae: *Ceciliodes acicula*.

Cecilioides acicula (Müller) (Fig. 13)
Thin, transparent, colourless shell, tapering very slowly to a blunt spire. The mouth is long and thin, about ⅓ the height of the shell. Height 4–5 mm, width 1·0–1·3 mm. A subterranean species, found in calcareous soils, usually in grassland. Found throughout England and Wales, S. and Central Ireland, but apparently absent from Scotland and N. Ireland.

Family CLAUSILIIDAE (all sinistral)

Marpessa laminata (Montagu) (Fig. 14a)

Long, glossy translucent shell, with very faint striations, usually pale brown, (not infrequently white). Mouth wide, with a thin white lip and conspicuous parietal and columellar folds. Height 16–18 mm, width about 4 mm. A characteristic woodland species, found also in hedges, or near isolated trees. Common throughout England, but only locally scattered in Ireland and Scotland.

Clausilia bidentata (Strom) (Fig. 14b)
(*C. rugosa* Draparnaud)

Long dull brown shell with faint white vertical streaks, moderately striated, with a small mouth. Height very variable, 8–13 mm, but width never as much as 3 mm (usually about 2·5 mm). A very widespread species, common in woods, hedges, rocks and walls, throughout the British Isles.

Clausilia dubia Draparnaud (Fig. 14c)

A larger, but otherwise similar species to *C. bidentata*, lacking the white streaks (but the worn surfaces of striations show pale grey). Height 15–17 mm, width 3·5 mm. Lives on rocks and screes on limestone, in the northern Pennines, the outer Hebrides and also at Dover (introduced).

Clausilia rolphii Leach (Fig. 14d)

Shell much wider in proportion to length than other *Clausilias*, thinner, lighter brown, and more heavily striated. The mouth is much wider than in *C. bidentata*. Height 11–13 mm, width 3·0–3·5 mm. Restricted to calcareous woods and hedges in southern and central England.

Lacinaria biplicata (Montagu) (Fig. 14e)

Large, wide shell with very heavy striations, and pale brown. Shell similar in height to *C. dubia*, but shell and mouth wider, and more coarsely striated. Height 16–17 mm, width 4 mm. Usually near river-banks on willows. Restricted to the London area, Hertfordshire and Cambridge.

Balea perversa (L.) (Fig. 14f)

The smallest member of the family, light brown, irregularly striated shell, widest at the body whorl, rather than higher up, as in the others. Mouth rounded and simple, sometimes with a single parietal tooth. Height 7–9 mm, width 2–2·5 mm. Found on trees, rocks and walls, only rarely on the ground, in a wide variety of habitats, but never in large numbers. Found throughout the British Isles.

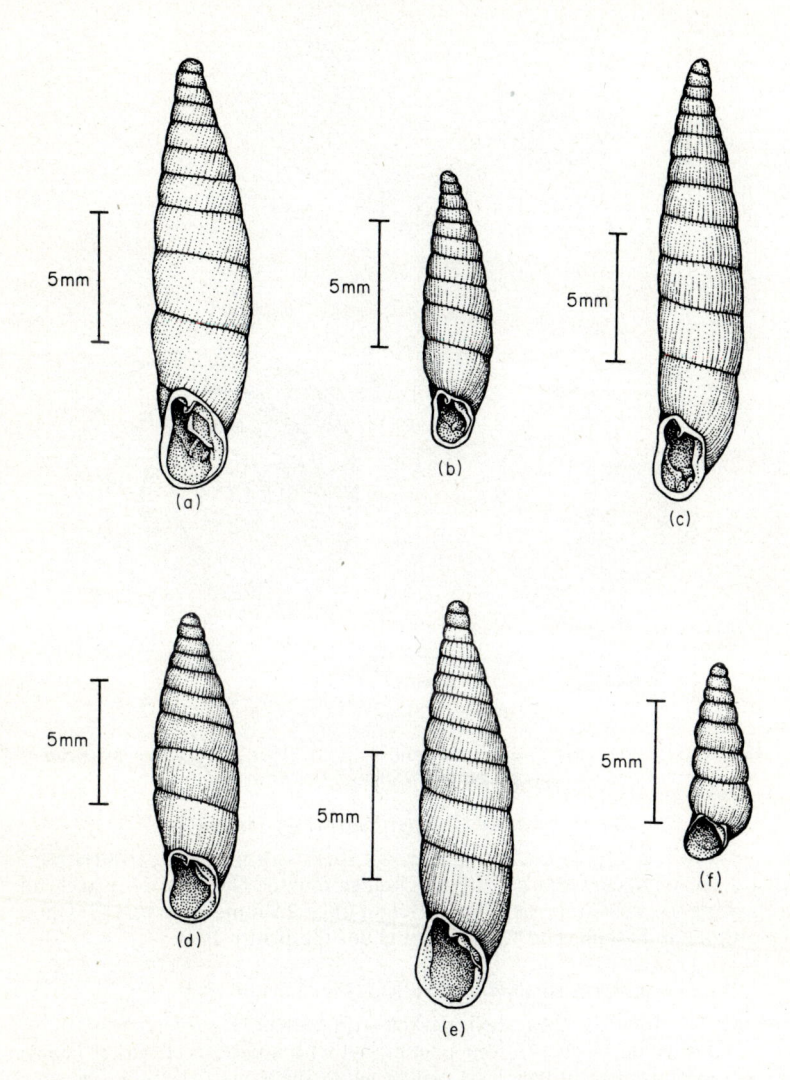

FIG. 14. Clausiliidae: (a) *Marpessa laminata*, (b) *Clausilia bidentata*, (c) *C. dubia*, (d) *C. rolphii*, (e) *Lacinaria biplicata*, (f) *Balea perversa*.

Family ENDODONTIDAE

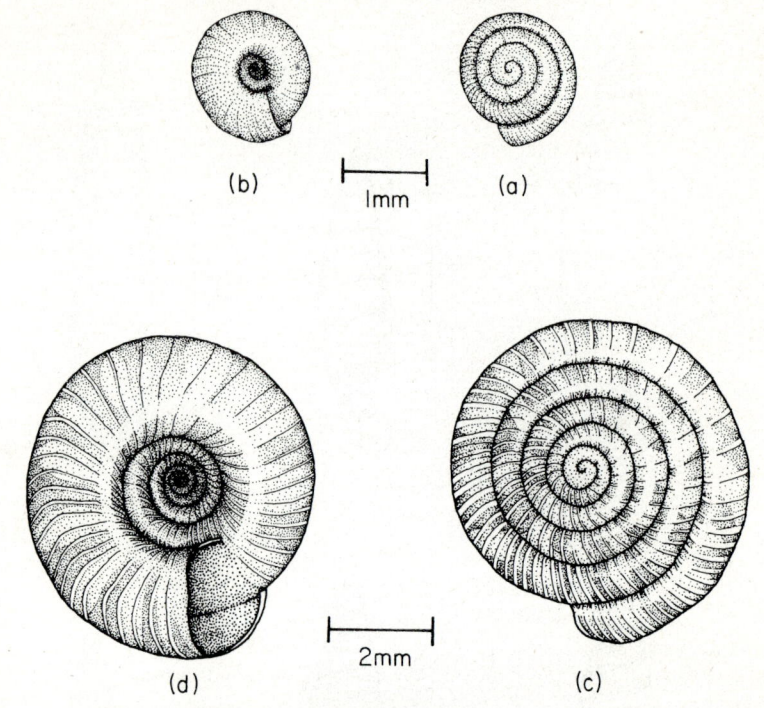

FIG. 15. Endodontidae: (a) and (b) top and bottom views of *Punctum pygmaeum*; (c) and (d) top and bottom views of *Discus rotundatus*.

Punctum pygmaeum (Draparnaud) (Fig. 15a and b)

Minute, flattened, tightly coiled light brown shell with fine, regular striations. Large, deep umbilicus, and an almost circular mouth. No trace of a keel, all whorls being rounded at the periphery. Height up to 0·9 mm, width up to 1·7 mm. Very catholic in habitat, and found throughout the British Isles.

Discus rotundatus (Müller) (Fig. 15c and d)

Flattened, but slightly conical shell marked with distinct large regular striations, and with a huge umbilicus. Colour pale brown with a pattern of darker brown blotches on the upper surface (completely white shells may be found in some populations). The edges of the whorls are slightly keeled or angulate, and the mouth is much wider than its height. Height 2·5–3·0 mm, width 6·0–7·5 mm. Found typically in woodland, in litter and on rotting wood, but widespread in other habitats also, being absent only from extremely dry habitats. Common throughout the British Isles.

Family ZONITIDAE

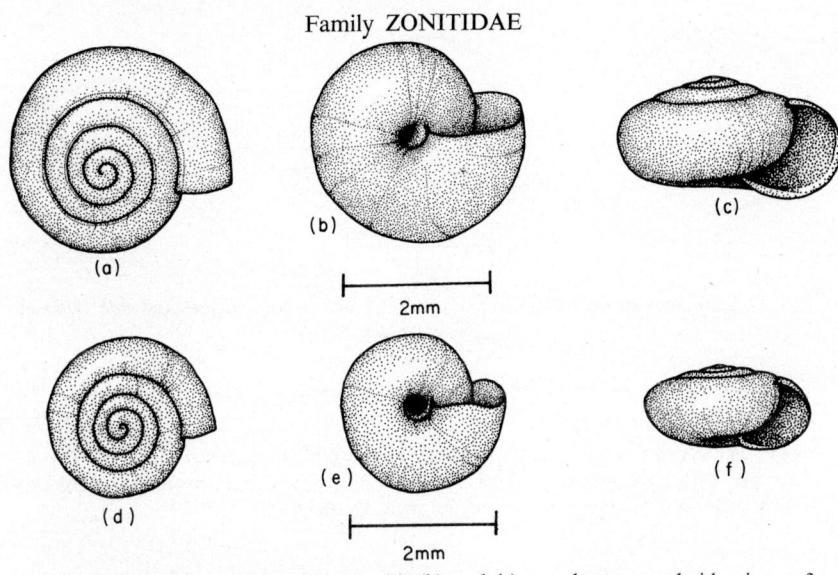

FIG. 16. Zonitidae genus *Vitrea* in part. (a), (b) and (c) top, bottom and side views of *V. crystallina*; (d), (e) and (f) top, bottom and side views of *V. contracta*.

Vitrea crystallina (Müller) (Fig. 16a, b and c)

Tightly coiled, thin, transparent shell, with small and noticeably eccentric umbilicus. The width of the last whorl at the mouth is more than $1\frac{1}{2} \times$ the width of the next, the mouth is rounded, not flattened below, and may have a slight columellar rib. Height 1·5–1·7 mm, width 2·8–3·3 mm. Found in a wide range of habitats, but able to tolerate wetter and more acid situations than *V. contracta*, with which, however, it is frequently found. Generally distributed in the British Isles.

Vitrea contracta (Westerlund) (Fig. 16d, e and f)

Tightly coiled, thin translucent-whitish shell, with a deep and circular umbilicus. Width of last whorl less than $1\frac{1}{2} \times$ width of next at the mouth. The mouth, and shell generally are more flattened on the under surface, and there is no columellar rib. Height 1·2–1·3 mm, width 2·0–2·5 mm. Like *V. crystallina*, it is found in a wide range of habitats, but with a preference for calcareous and dry sites. Generally distributed in the British Isles.

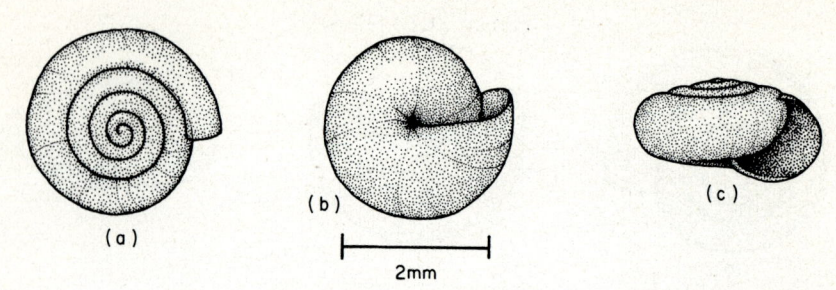

FIG. 17. Zonitidae, genus *Vitrea* (in part): (a), (b) and (c) top, bottom and side views of *V. diaphana*.

Vitrea diaphana (Studer) (Fig. 17a, b and c)
(*V. subrimata* Reinhardt)
Shell very similar in general appearance to those of *V. contracta* and *V. crystallina*, but with a minute umbilicus. Height 1·2 mm, width 2·3 mm. Restricted to rocky ground and screes on limestone in N.W. England.

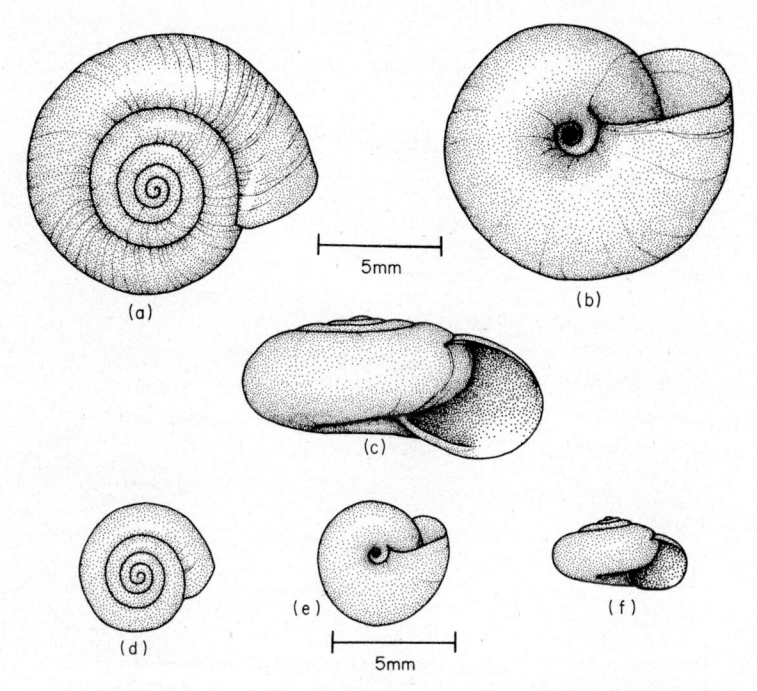

FIG. 18. Zonitidae, genus *Oxychilus* (in part): (a), (b) and (c) top, bottom and side views of *O. draparnaldi*, (d), (e) and (f) top, bottom and side views of *O. alliarius*.

Oxychilus draparnaldi Beck (Fig. 18*a*, *b* and *c*) K73
(*O. lucidus* (Draparnaud))

Shell very flattened, pale brown, glossy, but with moderate and irregular striations, especially near the mouth. The last whorl enlarges more rapidly near the mouth, which is set obliquely to the horizontal axis of the shell. The body is dark blue-grey, including the sole. Height 6–7 mm, width 13–15 mm. Found predominantly in gardens, but also in woods in S.W. England — the association with man becomes more marked in the north, but this species is widely distributed in the British Isles.

Oxychilus alliarius (Miller) (Fig. 18*d*, *e* and *f*) K74

Shell less flattened, very glossy and brown, the umbilicus relatively large. The body is dark blue-black. Height 2·5–3·5 mm, width 6–7 mm. Apart from its size, it may be distinguished from all other *Oxychilus* spp. when alive by the strong smell of garlic given off when prodded. The shell most resembles that of a young *O. helveticus*, but the umbilicus is proportionately larger. Very catholic in habitat, and more tolerant of acid conditions than *O. cellarius*. Found throughout the British Isles.

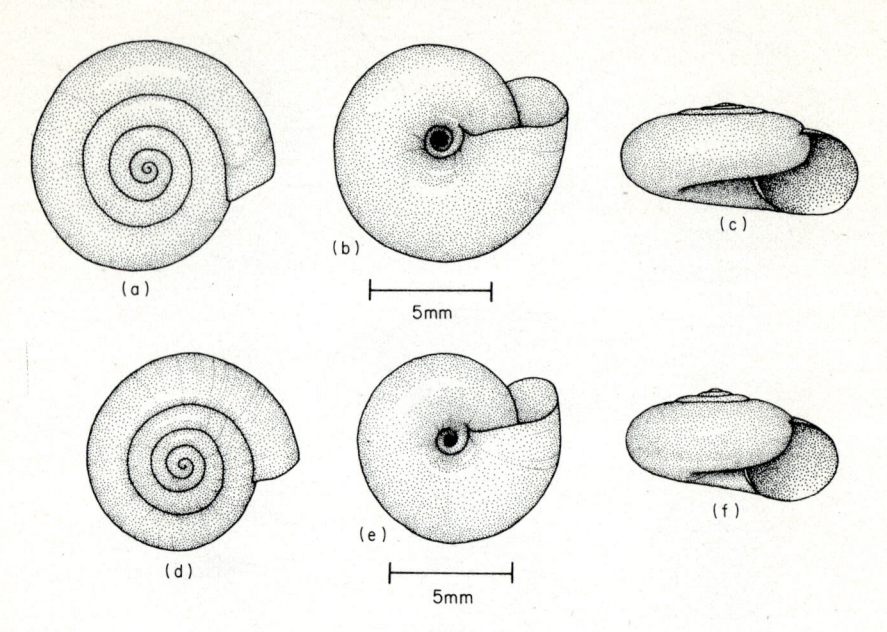

FIG. 19. Zonitidae, genus *Oxychilus* (in part): (a), (b) and (c) top, bottom and side views of *O. cellarius*; (d), (e) and (f) top, bottom and side views of *O. helveticus*.

Oxychilus cellarius (Müller) (Fig. 19a, b and c)

Shell very flattened, glossy and transparent, very pale brown, scarcely at all striated. Body and mantle pale grey, the latter with brown spots, and a pale sole. Height 5 mm, width 10 mm. Apart from size, it may be distinguished from *O. draparnaldi* by its extreme smoothness and glossiness, by the less obliquely set mouth, narrower last whorl at the mouth, and by body colour. Distinguished from *O. helveticus* by paler shell colour, flatter spire, relatively larger umbilicus and body colour. A common woodland species, but occurring in a wide range of other habitats, including gardens. Found throughout the British Isles.

Oxychilus helveticus (Blum) (Fig. 19d, e and f)

Shell less flattened, glossy and brown, with a smaller umbilicus than in other *Oxychilus* spp. Height 4–5 mm, width 9–10 mm. The body is blue grey, and the mantle is very dark, and contrasts with the rest of the contents of the shell. A woodland and hedgerow species, but less common and more localized than *O. cellarius* and *O. alliarius*. Throughout England and Wales, but scarce in Scotland.

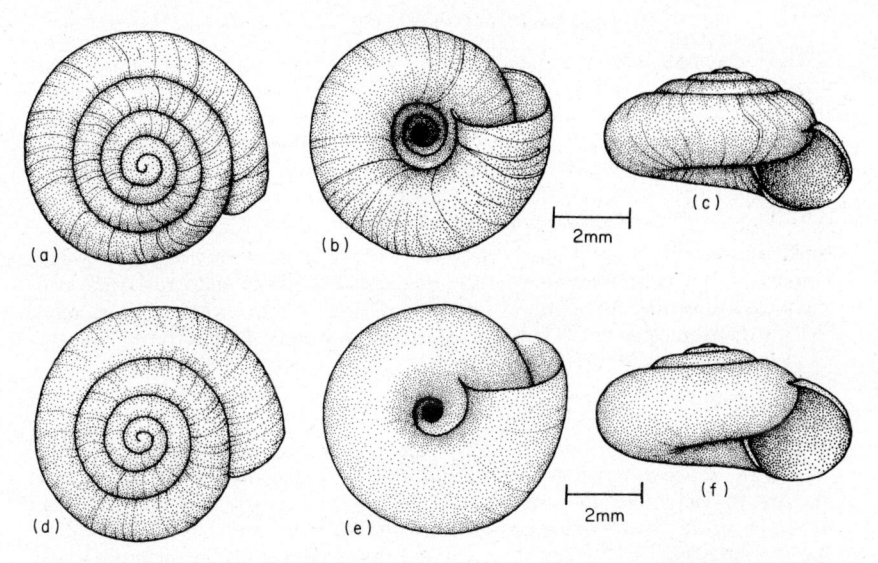

FIG. 20. Zonitidae, genus *Zonitoides*: (a), (b) and (c) top, bottom and side views of *Z. excavatus*; (d), (e) and (f) top, bottom and side views of *Z. nitidus*.

Zonitoides excavatus (Alder) (Fig. 20a, b and c)

Shell glossy, brown and tightly coiled, with a very large umbilicus ($\frac{1}{4}$ or more width of whole shell). Height 3 mm, width 6–7 mm. The mantle is grey with white spots. The size of the umbilicus and spotting on the mantle distinguish it from *Z. nitidus*. A woodland species, apparently confined to acid soils. Widely distributed in the British Isles, but rare or absent in many eastern counties of Great Britain and Ireland.

Zonitoides nitidus (Müller) (Fig. 20d, e and f)

Shell glossy, brown, not so tightly coiled as *Z. excavatus*, with a smaller umbilicus ($\frac{1}{5}$ width of shell). Height 4 mm, width 8 mm. The mantle is grey with black spots and a single orange spot on the right. The live snail appears very dark due to the black body shewing through the shell. Characteristic of very wet habitats, never far from water, and may occasionally be found immersed. Widespread in the British Isles, but becoming rare in northern Scotland.

Retinella radiatula (Alder) (Fig. 21*a*, *b* and *c*)

Shell transparent, glossy and pale brown, with rapidly expanding whorls, and with very regular widely spaced striations, and a large but rather eccentric umbilicus. Height 2 mm, width 4 mm. One of the most widely distributed British snails occurring in all but very dry, calcareous habitats, and quite tolerant of acidity. Found throughout the British Isles.

Retinella pura (Alder) (Fig. 21*d*, *e* and *f*)

Shell translucent, slightly glossy with faint irregular striations. Usually white/colourless, but brown shells are quite frequent. Umbilicus wide and deep, and much less eccentric than in *R. radiatula*. Height 2 mm, width 4·0–4·5 mm. Widely distributed in damp habitats, especially in woods. Found throughout the British Isles.

Retinella nitidula (Draparnaud) (Fig. 21*g*, *h* and *i*)

Shell only slightly translucent, waxy, pale brown, becoming white round the umbilicus, which is large. Whorls expand rapidly, especially the last whorl near the mouth. Height 4 mm, width usually 7–8 mm but occasionally up to 11 mm. Juveniles of this species may be confused with *R. pura*; they have fewer whorls for the same size, a larger protoconch and the striations on the shell are much coarser. Doubtful specimens may be compared with the spire of an adult *R. nitidula*. In woods and hedges throughout the British Isles.

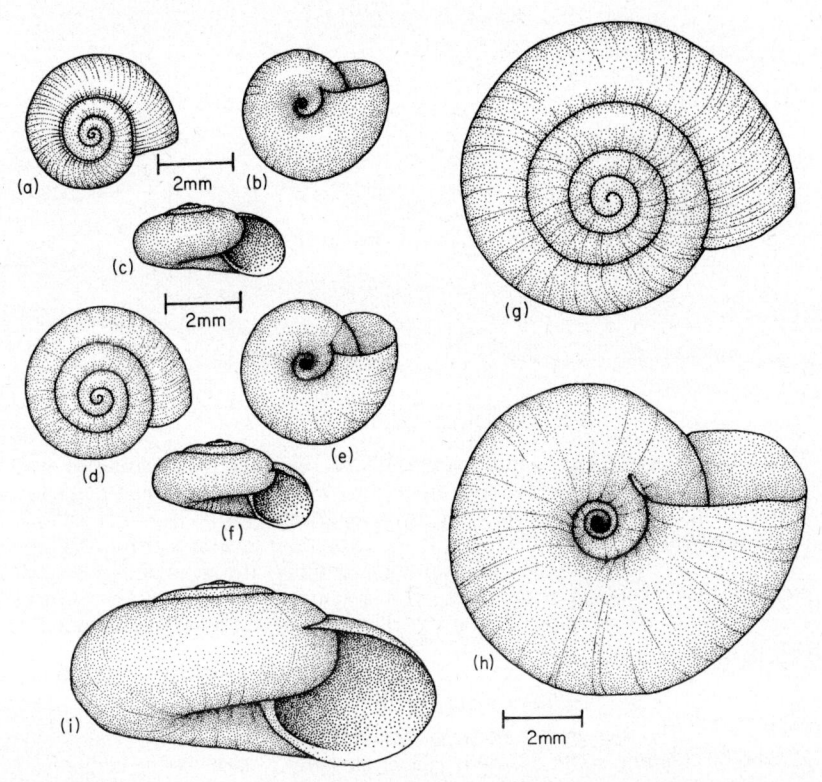

FIG. 21. Zonitidae genera *Retinella*: (a), (b) and (c) top, bottom and side views of *Retinella radiatula*; (d), (e) and (f) top, bottom and side views of *R. pura*; (g), (h) and (i) top, bottom and side views of *R. nitidula*.

Family VITRINIDAE

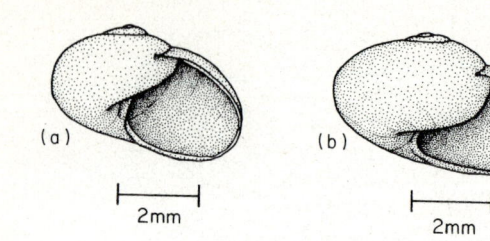

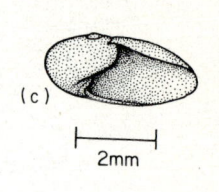

FIG. 22. Vitrinidae, genus *Vitrina*: (a) *V. pellucida*, (b) *V. major*, (c) *V. pyrenaica*.

Vitrina pellucida (Müller) (Fig. 22a)

Shell globular, very glossy and transparent, with very rapidly increasing whorls. Height 3 mm, width 5–6 mm. Body light grey, with darker head and tentacles. Living adults found in winter months. By comparison with *V. major*, the spire is taller, the mouth is rounder, and the body colour and time of activity of adults differ, but certainty can be obtained by examining the genitalia—the spermathecal duct, penis and oviduct join together at approximately the same place in *V. pellucida*. Very widespread and catholic in habitat; found throughout the British Isles.

Vitrina major Férussac (Fig. 22b)

Shell very similar to that of *V. pellucida*, but flatter, and with a more oval mouth. Height 3·0–3·5 mm, width 6–7 mm. The whole body is dark. Living adults tend to be found in summer. The spermathecal duct joins the oviduct just before the latter swells into a characteristic gland, and considerably above the junction of the oviduct and penis. Found mostly in woodland. Local in distribution, occurring in scattered localities in southern and western England and S. Wales.

Vitrina pyrenaica Férussac (Fig. 22c)

Shell similar in transparency and glossiness to the preceding two species, but with whorls expanding much more rapidly, with a huge oval mouth. Height about 2 mm, width about 5 mm. When active, almost the whole of the shell is covered by folds of the mantle. Known only from a few localities in southern Ireland.

Family EUCONULIDAE

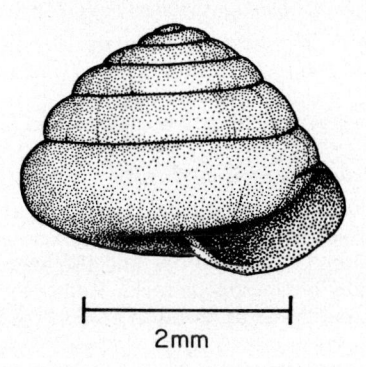

FIG. 23. Euconulidae, genus *Euconulus*: side view of *E. fulvus*.

Euconulus fulvus (Müller) (Fig. 23)

Shell tightly coiled, slightly glossy, translucent and brown, with a conical spire and a minute umbilicus. Height 2·0–2·5 mm, width 3·0–3·5 mm. Found in many habitats, especially in woods and wetlands, but rare or absent in very dry habitats. Found throughout the British Isles.

Family BRADYBAENIDAE

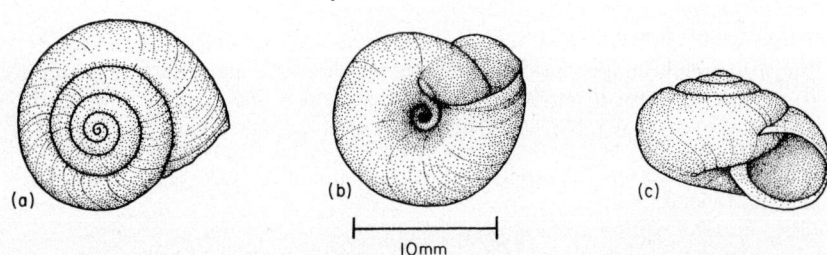

(a) (b) (c)

FIG. 24. Bradybaenidae: top, bottom and side views of *Fruticicola fruticum*.

Fruticicola fruticum (Müller) (Fig. 24*a*, *b* and *c*)

(*Eulota fruticum* (Müller))

Shell globular, yellowish-white, with slight regular striations. The mouth is rounded, and there is a white rib behind the mouth edge. Height 16–17 mm, width 18–20 mm. The body is yellowish. Found only in a few hedgerows in Kent, almost certainly introduced.

Family HELICIDAE

Helicodonta obvoluta (Müller) (Fig. 25a, b and c)

Shell very flattened, with spire actually below the level of the body whorl. The mouth is thickened and triangular. The shell is brown, and covered with hairs, which are often worn off adult specimens. Height 5 mm, width 10–12 mm. Restricted to calcareous woodland in West Sussex and S.E. Hampshire.

Helicigona lapicida (L.) (Fig. 25d and e)

Shell rather flattened, with a very low conical spire and a very pronounced keel. Shell dull brown in colour, with a pattern of darker radial blotches. Height 7–8 mm, width 16–19 mm. Found in woods, and also in man-made and naturally rocky habitats, especially on calcareous rocks. Rather localized in distribution, but found in England and Wales as far north as Yorkshire, also at Cork.

Arianta arbustorum (L.) (Fig. 25f and g)

Shell large and globular, with a white lip reflected over the umbilicus, which is in consequence nearly completely closed. The shell is polymorphic; the commonest form is brown with a single dark band, but unbanded and yellow shells are also found. The shell is nearly always flecked with lighter patches. Shell size rather variable, typically height 16–18 mm, width 19–22 mm. Generally found in damp, cool habitats, but not in marshes. Widely distributed in Great Britain (rather rare in the extreme S.W.) but absent from the southern two thirds of Ireland.

Theba pisana (Müller) (Fig. 25h and i)

(*Euparypha pisana* (Müller))

Thick, white, globular shell, usually with many interrupted dark spiral bands, and with a pinkish flush inside the mouth. Height 14–16 mm, width 18–20 mm. By comparison with *Helicella virgata*, the umbilicus is smaller, the spire flatter, and the maximum size greater, besides the colouring inside the mouth. Juveniles are keeled, and may be mistaken for other keeled species. Restricted to sand-dunes in a few localities in Channel Islands, Cornwall and S. Wales, and in the east coast of Ireland.

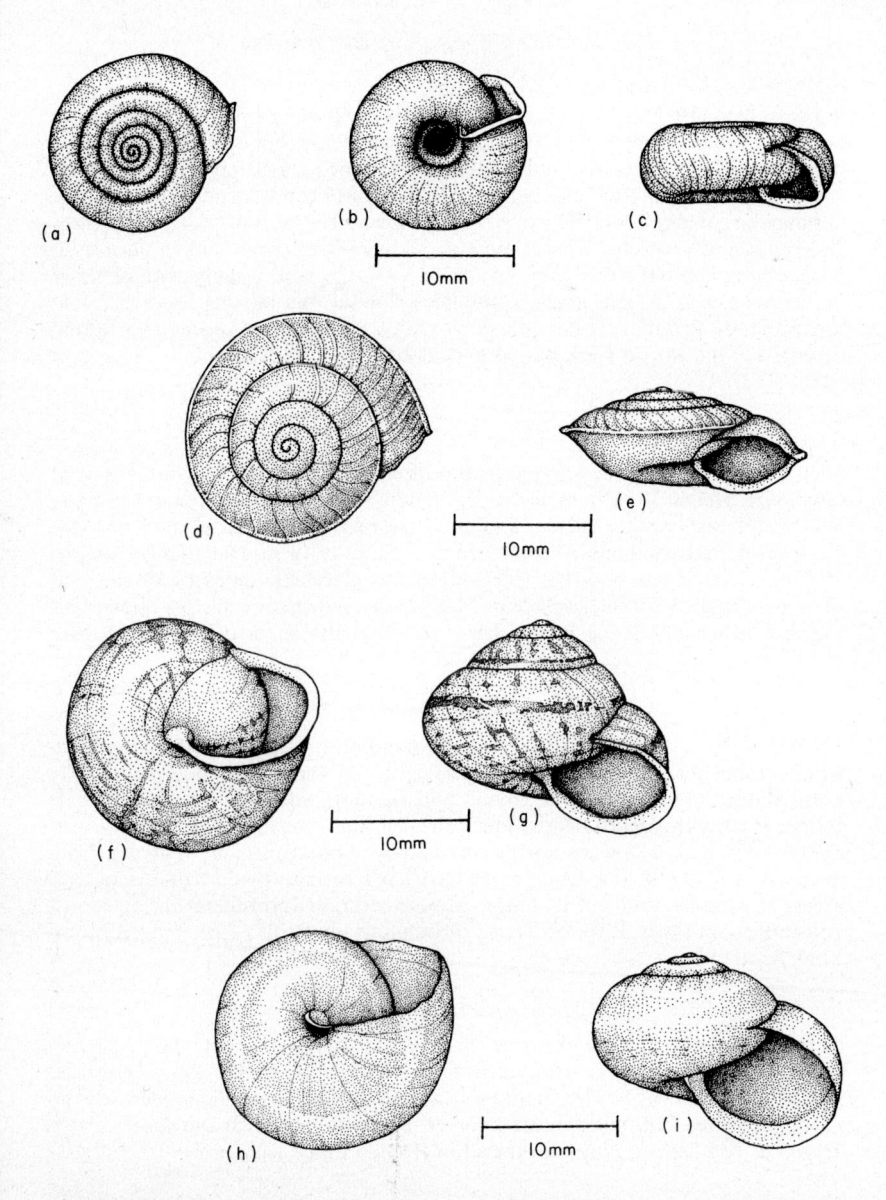

FIG. 25. Helicidae (in part): (a), (b) and (c) top, bottom and side views of *Helicodonta obvoluta*; (d) and (e) top and side views of *Helicigona lapicida*; (f) and (g) bottom and side views of *Ariarta arbustorum*; (h) and (i) bottom and side views of *Theba pisana*.

Helix hortensis Müller (Fig. 26*a*, *b* and *c*)

(*Cepaea hortensis* (Müller))

Brightly coloured globular shell, with a thickened white lip which completely seals the umbilicus. Height 14–16 mm, width 16–20 mm (but occasionally outside this range). The species is very polymorphic for shell colour and banding. Shell coloured yellow, pink or brown, and with 0–5 spiral bands, usually dark brown, but may be paler or completely devoid of pigment. The lip is occasionally brown, and is not always a reliable discriminator between this species and *H. nemoralis*. *H. hortensis* has bifurcated ends to the dart-flanges, and usually four or more branches to each mucous gland. Catholic in habitat, but missing from very acid sites, and from dunes within the geographical range of *H. nemoralis*. Found throughout the British Isles, but very local and rare in S.W. Ireland.

Helix nemoralis L (Fig. 26*d*, *e* and *f*)

(*Cepaea nemoralis* (L))

Shell similar to that of *H. hortensis*, but usually with a dark brown lip, and of larger size. Height 14–20 mm, width 18–26 mm, the largest shells found mostly in W. Ireland. Shell polymorphism similar to that of *H. hortensis*, but more variants are known, and populations tend to contain more of them. Dart-flanges simple, and usually 3 or less branches to each mucous gland. Occupies a wide range of habitats, usually in drier places than *H. hortensis*, but the two species often occur together. Throughout the British Isles, except in the Scottish Highlands north of Skye.

Helix aspersa Müller (Fig. 26*g*)

Large rough-surfaced shell with rapidly expanding whorls and a thick white lip which completely seals the umbilicus. Height and width very variable, usually both about 30–35 mm. Shell colour and banding variable, but usually pale brown shell with darker bands, much flecked and interrupted with patches of lighter colour. A yellow variant occurs locally. Found in many habitats, from dunes to woodland, but tends to be restricted to man-made habitats in acid areas, or near the limits of its range. Almost certainly introduced during or just before Roman times, it is now found throughout the British Isles, but very local in N. Scotland.

Helix pomatia L. (Fig. 26*h*)

Very large shell, thick, whitish or creamy brown, usually with faint, slightly darker bands. The lip is reflected over the umbilicus, which is not, however, completely closed by it. Height and width both 45 mm. Found in moderately dry calcareous sites, but in very local colonies, in central southern and south eastern England. Almost certainly introduced in Roman times.

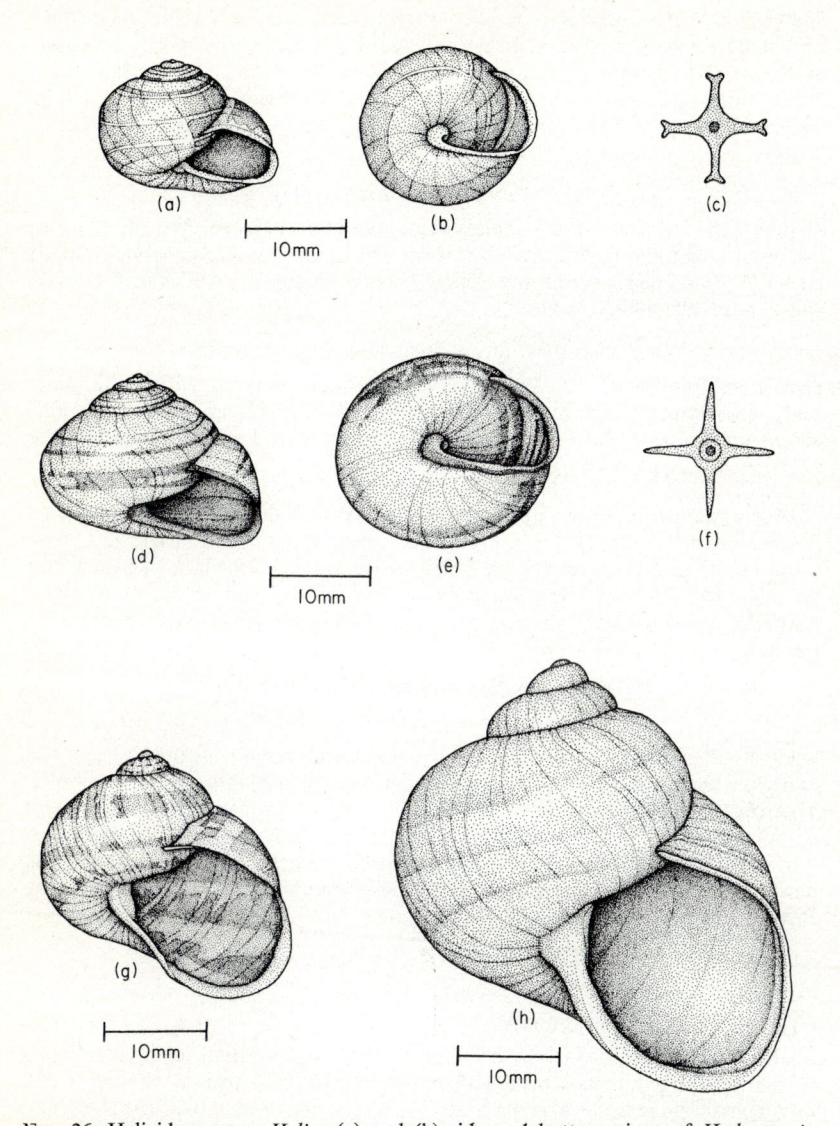

FIG. 26. Helicidae, genus *Helix*: (a) and (b) side and bottom views of *H. hortensis*, (c) cross-section of dart of *H. hortensis*, dissected out of the dart sac; (d) and (e) side and bottom views of *H. nemoralis*; (f) cross-section of dart of *H. nemoralis*, dissected out of the dart sac; (g) *H. aspersa*; (h) *H. pomatia*.

Hygromia limbata (Draparnaud) (Fig. 27*a*, *b* and *c*) K80

Shell slightly glossy, globular, and very slightly keeled; there is a thickened white rib behind the mouth edge, and the umbilicus is very small and partly obscured by the reflected mouth edge. Colour dark brown to yellow, often with a single spiral band which may be darker or lighter than the rest of the shell. Found in hedges and scrub in a few localities in S. Devon, and recently also near Malvern, Worcs.

Hygromia cinctella (Draparnaud) (Fig. 27*d*) K42

Slightly flattened shell with a conical spire, and a marked keel. No rib or lip at the mouth, but the edge is reflected over the umbilicus, which is very small. Height 6–7 mm, width 10–12 mm. Known only from a few localities in S. Devon, and almost certainly introduced.

Hygromia subvirescens (Bellamy) (Fig. 27*e* and *f*) K46

Shell thin, translucent and greenish, very globular, covered with white hairs. Body whorl much wider than the rest. Umbilicus small. Height 4–5 mm, width about 6 mm. Restricted to sites near the coast in S.W. England and S. Wales.

Hygromia subrufescens (Miller) (Fig. 27*g*, *h* and *i*) K69

Shell glossy and thin, brown in colour, with a very small umbilicus partly obscured by the columellar lip. Height about 6 mm, width 9–10 mm. Flatter, and more transparent than *H. limbata*, from which it also differs by the lack of bands, and the much less developed lip. Found mostly in old woodland, in the western and northern parts of Great Britain, and in central and mountainous districts of Ireland.

Hygromia striolata (Pfeiffer) (Fig. 27*j*, *k* and *l*) K83

(*Trichia striolata* (Pfeiffer))

Shell slightly flattened above, and with a slight keel. Umbilicus open. There is a marked white rib just behind the mouth edge. Height 7–10 mm, width 11–14 mm. There is a shell polymorphism, the colour varying from white to dark brown, and a faint pale spiral band is often present at the keel. Juveniles are hairy, but much more obviously keeled, and thus distinguishable from *H. hispida*. Found in most damp habitats—hedges, gardens, roadsides—also in woods, throughout the British Isles.

Hygromia hispida (L) (Fig. 27*m*, *n* and *o*) K45

(*Trichia hispida* (L))

(*Hygromia liberta* (Westerlund))

Slightly flattened shell, lacking any keel, with an open umbilicus. Shell covered with short curved hairs which may be worn off (a few usually remain in the umbilicus). There is a white rib strengthening the mouth. Height 4–6 mm, width 5–11 mm. The shell has a colour polymorphism similar to that of *H. striolata*, but brown shells are much the commonest. There is much variation in shell size and shape, and some variations have been regarded as species (e.g. *H. liberta*). Found in a very wide range of habitats, from grassland to woods and wetlands, throughout the British Isles, but rare and localized in N. Scotland.

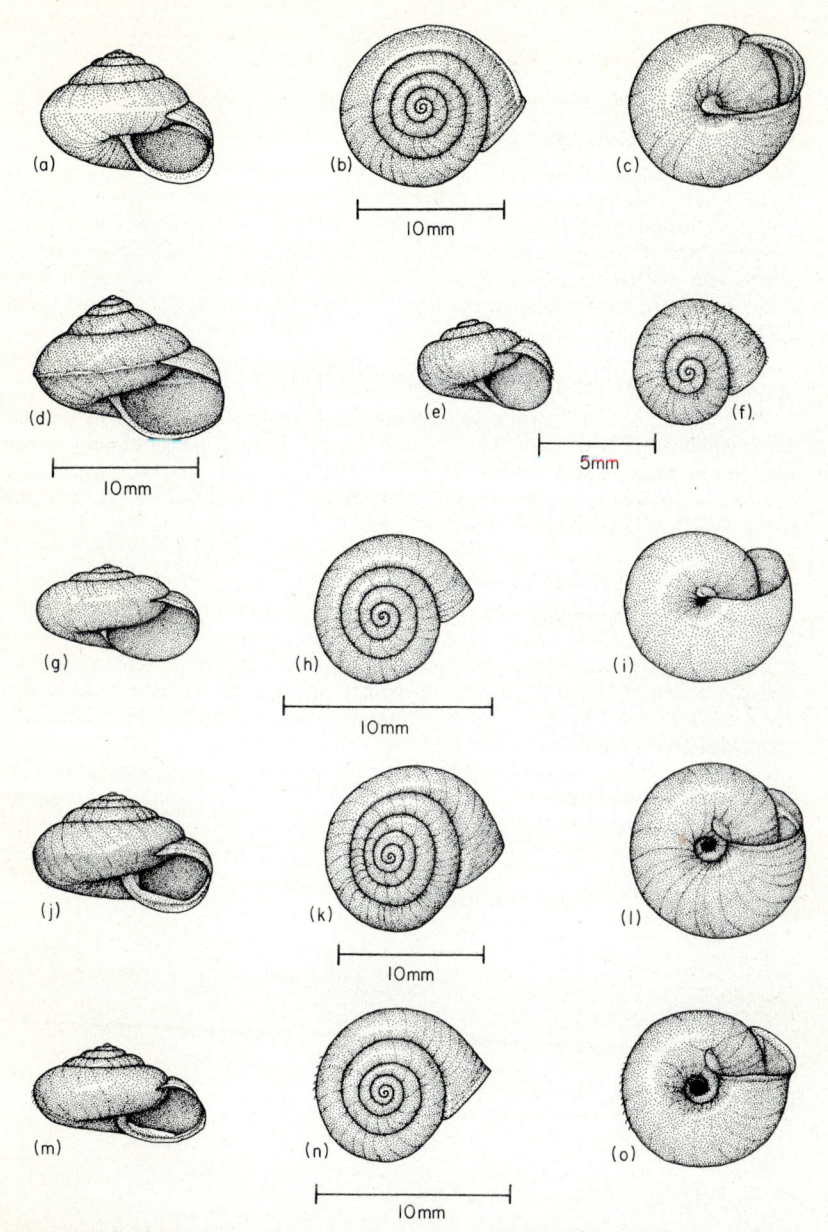

FIG. 27. Helicidae, genus *Hygromia*: (a), (b) and (c) side, top and bottom views of *H. limbata*; (d) *H. cinctella*; (e) and (f) side and top views of *H. subvirescens*; (g), (h) and (i) side, top and bottom views of *H. subrufescens*; (j), (k) and (l) side, top and bottom views of *H. striolata*; (m), (n) and (o) side, top and bottom views of *H. hispida*.

Monacha granulata (Alder) (Fig. 28a, b and c) K46
(*Ashfordia granulata* (Alder))
Shell thin, translucent and hairy, the hairs being long and straight with bulbous bases. Umbilicus very small, and partially covered by reflected mouth edge. Colour yellowish-grey. Height 5–6 mm, width 6–8 mm. The straight hairs on the shell and size of umbilicus serve to distinguish it from *Hygromia hispida*, and by comparison with *H. subvirescens* it is larger, and has a much more prominent spire. Found in wet places, in herbage, in England and Wales, but is rare in Scotland and Ireland.

Monacha cartusiana (Müller) (Fig. 28d, e and f) K80
Shell globular, flattened, white and opaque, with a white rib behind the mouth-edge, which is brown. Umbilicus minute. Height 8 mm, width 12 mm. Some varieties lack the brown mouth-edge. Found in calcareous grasslands, but rare and local, Suffolk, Kent, Sussex and Hampshire only, and always fairly near the coast. Possibly introduced in Neolithic times.

Monacha cantiana (Montagu) (Fig. 28g, h and i) K85
Shell globular, but slightly flattened, thin and translucent, and whitish, but usually with a reddish brown flush near the mouth, which fades rapidly away from the mouth. A pale opaque, central spiral band may be present. Umbilicus about $\frac{1}{5}$ width of last whorl opposite mouth. Small juveniles have hairy shells. Height 10–12 mm, width 15–18 mm. In dry, herbaceous habitats, especially roadside verges; common in south, central and eastern England up to Yorkshire, becoming very rare and local in S.W. England and Wales, and absent from N.W. England, N. Wales, Ireland and Scotland (except for one introduced population). Probably introduced in Roman times.

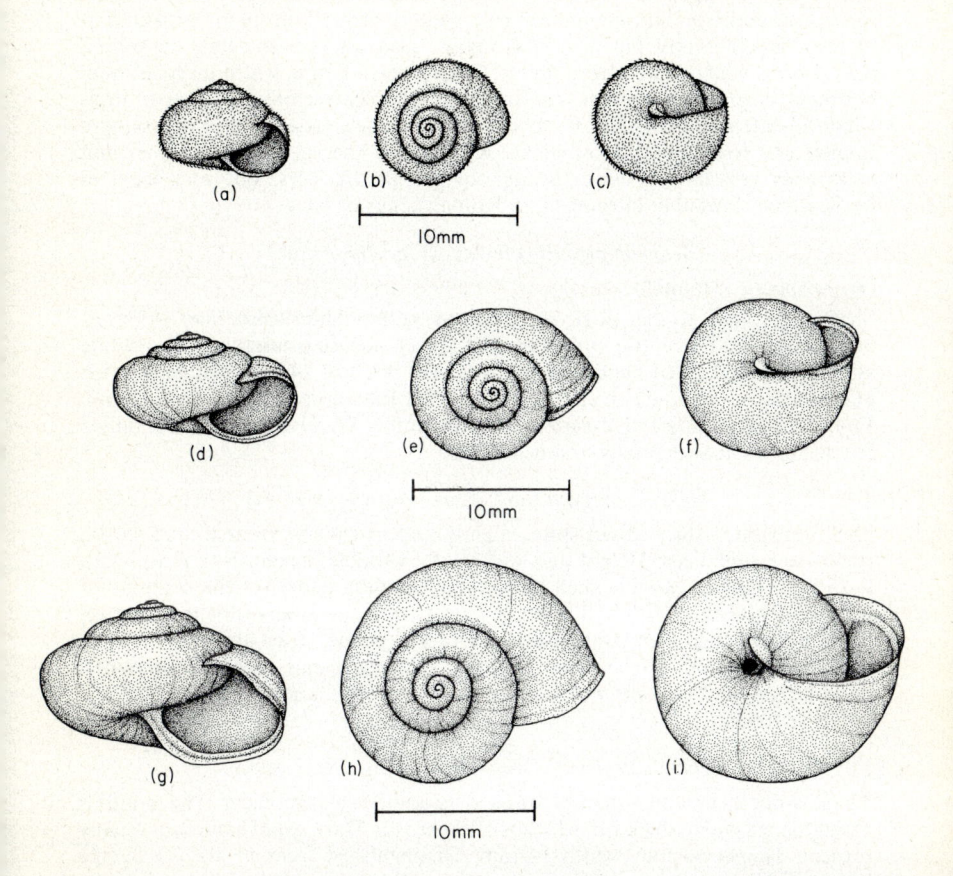

FIG. 28. Helicidae, genus *Monacha*: (a), (b) and (c) side, top and bottom views of *Monacha granulata*; (d), (e) and (f) side, top and bottom views of *M. cartusiana*; (g), (h) and (i) side, top and bottom views of *M. cantiana*.

Helicella caperata (Montagu) (Fig. 29*a*, *b* and *c*)

Shell thick and opaque, with heavy striations. The shell is highly polymorphic for colour and banding; commonly pale gingery brown with many broken and blotchy darker bands, but not uncommonly also white, with single dark peripheral band when viewed from above. Height up to 8 mm, width up to 12 mm, but many become adult much smaller. It most closely resembles *H. gigaxii*, from which it differs in being less flattened, more coarsely striated, and in having a smaller and less eccentric umbilicus. A xerophile species, found in dunes and calcareous grassland. Found throughout the British Isles, but very local in N. Scotland. Probably introduced in Roman times or later.

Helicella gigaxii (Pfeiffer) (Fig. 29*d*, *e* and *f*)
(*H. heripensis* (Mabille))

Shell very similar to that of *H. caperata*, but with a flatter spire, finer striations and a larger and eccentric umbilicus. The upper surface is usually creamy-white with faint, interrupted bands. Height 4–7 mm, width 8–14 mm. Found in open and herbaceous habitats on calcareous soil, but less xerophile than *H. caperata*. Found locally in England, except in the S.W. and N.W., also in Wales, but only a few localities in S. Scotland and in Ireland.

Helicella virgata (da Costa) (Fig. 29*g*, *h* and *i*)

Shell globular, with a raised spire, slightly and irregularly striated, and with a small deep umbilicus. Height and width very variable, height 7–18 mm, width 8–25 mm. Highly variable shell colour and banding patterns—the commonest forms being a white shell with a single dark peripheral band visible from above, and a gingery shell with faint darker interrupted bands. Often abundant in open habitats throughout England and Wales on calcareous soils, especially on dunes, rarer in Scotland and absent from the extreme north. Local, but widely distributed in Ireland.

Helicella neglecta (Draparnaud) (Fig. 29*j*, *k* and *l*)

Thick white shell, with a slightly raised spire and a large umbilicus. The mouth is strengthened with a thick rib, which is often brown. Dark spiral bands are usually present. Height 10 mm, width 14 mm. Distinguished from *H. virgata* by the larger umbilicus, and from *H. itala* by the taller spire and thicker internal rib. Found only in one locality in Kent, and almost certainly a recent introduction; possibly now extinct in Britain.

Helicella itala (L) (Fig. 29*m*, *n* and *o*)

Shell very flattened, slightly irregularly striated, and with a very large open umbilicus. Height 5–12 mm, width 9–25 mm. The shell is polymorphic for colour and banding; the commonest form, in most of Britain, is a white shell with a single dark peripheral band seen from above. A xerophile species, found on dunes and open calcareous grassland. Found throughout the British Isles, but appears to have become extinct in many inland areas from which it was previously known.

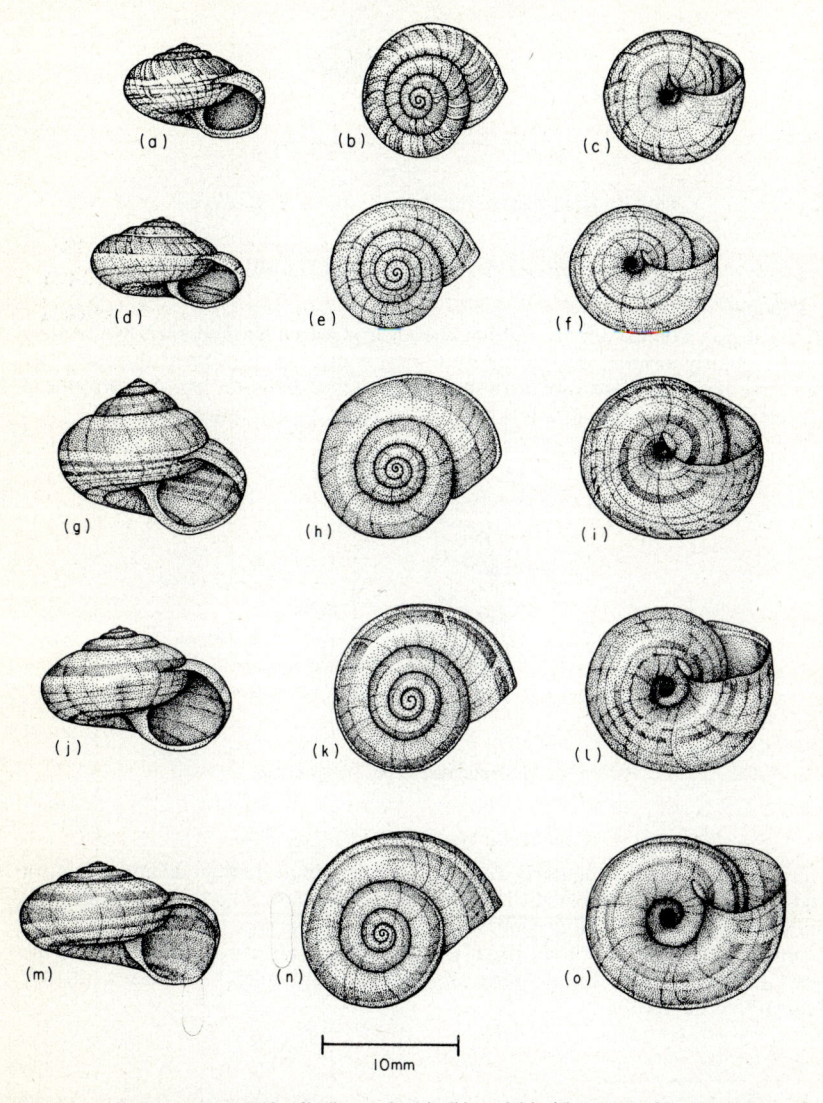

FIG. 29. Helicidae, genus *Helicella* (in part): (a), (b) and (c) side, top and bottom views of *H. caperata*; (d), (e) and (f) side, top and bottom views of *H. gigaxii*; (g), (h) and (i) side, top and bottom views of *H. virgata*; (j), (k) and (l) side, top and bottom views of *H. neglecta*; (m), (n) and (o) side, top and bottom views of *H. itala*.

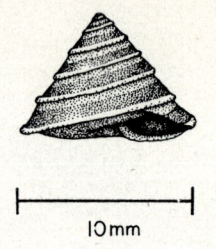

FIG. 30. Helicidae, genus *Helicelle* (in part): *H. elegans*.

Helicella elegans (Gmelin) (Fig. 30) K 42

(*Trochoidea elegans* (Gmelin))
Shell steeply conical above, and flattened below with a marked keel. Moderately and regularly striated. Height about 6 mm, width 7–8 mm. Shell shows colour and banding variations similar to that seen in other *Helicella* spp. An introduced species, now occurring in a few colonies in Kent, Sussex and Surrey.

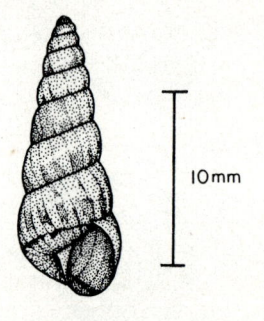

FIG. 31. Helicidae, *Cochlicella actua*.

Cochlicella acuta (Müller) (Fig. 31) K 38

Shell spire-shaped, irregularly striated, with a minute umbilicus and a mouth edge without rib or lip. Height 15–25 mm, width 5–7 mm. There is much variation in shell colour and banding, similar to that of *Helicella* spp. In Great Britain, it is confined to coastal localities, usually sand-dunes, being absent from most of the east coast, but it is found both on the coast and inland in Ireland, in dry calcareous habitats.

Acknowledgements

The key to species in this synopsis is derived from a field key to land-snails constructed by Dr. J. J. D. Greenwood and R.A.D.C. That key, and many subsequent editions were used and criticised by many people, whose comments were essential to its improvement. The Earl of Cranbrook first suggested that the key should form the basis of a synopsis. Dr. J. E. Chatfield, Dr. M. P. Kerney and Mr. A. Norris have very kindly lent us specimens, and Dr. Kerney has given us great help over the status, distribution and nomenclature of many species.

References

The references below include all those mentioned in the text and also other general works including information on terrestrial snails. Detailed descriptions of most species, and references to original descriptions can be found in Ellis (1969), but recent references to the separation of critical species have been included.

The Conchological Society of Great Britain and Ireland is running a 10 km square survey of non-marine molluscs which is now ending. Full details are given in Kerney (1967). Provisional distribution maps for some species have been published in recent numbers of the *Journal of Conchology*, which also carries annual reports of additions to vice-county lists, thus keeping the last supplement to the census (Kerney 1966) up to date.

BOYCOTT, A. E. 1934. The habitats of land Mollusca in Britain. *J. Ecol.* **22**, 1–38.

ELLIS, A. E. 1951. Census of the distribution of British non-marine Mollusca, 7th edition, *J. Conch. Lond.* **23**, 171–244.

ELLIS, A. E. 1969. *British Snails* (revised edition). Oxford.

EVANS, J. G. 1973. *Land Snails in Archaeology*, Academic Press, London.

FORD, E. B. 1964. *Ecological Genetics*. London, Methuen.

FRETTER, V. 1968. Studies in the structure, physiology and ecology of molluscs. *Symp. Zool. Soc. London* **22**.

FRETTER, V. and GRAHAM, A. 1962. *British Prosobranch Molluscs*. Ray Society, London.

FRETTER, V. and PEAKE, J. 1975. *Pulmonates*. Academic Press, London, New York and San Francisco.

JANUS, H. 1965. *The young specialist looks at Land & Freshwater Mollusca*. Burke, London.

KERNEY, M. P. 1966. Supplement to the census of the distribution of British non-marine Mollusca. *J. Conch. Lond.* **25**, appendix.

KERNEY, M. P. 1967. Distribution mapping of land and freshwater Mollusca in the British Isles. *J. Conch. Lond.* **26**, 152–160.

KERNEY, M. P. and FOGAN, M. 1969. *Vitrea diaphana* (Studer) in Britain. *J. Conch. Lond.* **27**, 17–24.

KUIPER, J. G. J. 1964. On *Vitrea contracta* (Westerlund). *J. Conch. Lond.* **25**, 276–278.

LLOYD, D. C. 1970. The use of skin characters as an aid to the identification of the British species of *Oxychilus* (Fitzinger) (Mollusca, Pulmonata, Zonitidae). *J. Nat. Hist.* **4**, 531–534.

MACAN, T. T. 1969. Key to the British fresh and brackish water Gastropods. *Freshwater Biological Association Scientific Publication* 13. 3rd edition.

MORTON, J. E. 1967. *Molluscs*. Hutchinson, London.

NORRIS, A. and COLVILLE, B. 1974. Notes on the occurrence of *Vertigo angustior* Jeffreys in Great Britain. *J. Conch. Lond.* **28**, 141–154.

PURCHON, R. D. 1968. *The Biology of the Mollusca*. Oxford, Pergamon.

QUICK, H. E. 1949. Slugs (Mollusca) (Testacellidae, Arionidae, Limacidae) *Synopsis of the British Fauna No.* 8. Linnean Society of London.

QUICK, H. E. 1954. *Cochlicopa* in the British Isles. *Proc. malac. Soc. Lond.* **30**, 204–213.

TAYLOR, J. W. 1894–1921. *Monograph of the Land and Freshwater Mollusca of the British Isles*, 3 vols. + 3 parts unfinished. Taylor Bros., Leeds.

WATSON, H. and VERDCOURT, B. 1953. The two British species of *Carychium*. *J. Conch. Lond.* **23**, 306–324.

ZILCH, A. 1959. Gastropoda, Euthyneura. *Handb. Palaozool.* **6** (2). Berlin. Gebrüder Bormtraeger.

Index of Species

Synonyms are shown in roman, the correct generic and specific names in italics. Figures in bold type refer to illustrations.

Abida secale	32 **33**	*Hygromia cinctella*	54 **55**
Acanthinula aculeata	34 **35**	*Hygromia hispida*	54 **55**
Acanthinula lamellata	34 **35**	Hygromia liberta	54 **55**
Acicula fusca	22 **22**	*Hygromia limbata*	54 **55**
Acme fusca	22 **22**	*Hygromia striolata*	54 **55**
Arianta arbustorum	50 **51**	*Hygromia subrufescens*	54 **55**
Ashfordia granulata	56 **57**	*Hygromia subvirescens*	54 **55**
Azeca goodalli	26 **26**	*Lacinaria biplicata*	38 **39**
Balea perversa	38 **39**	*Lauria anglica*	32 **33**
Carychium minimum	23 **23**	*Lauria cylindracea*	32 **33**
Carychium tridentatum	23 **23**	*Marpessa laminata*	38 **39**
Catinella arenaria	24 **25**	*Monacha cantiana*	56 **57**
Cecilioides acicula	37 **37**	*Monacha cartusiana*	56 **57**
Cepaea hortensis	52 **53**	*Monacha granulata*	56 **57**
Cepaea nemoralis	52 **53**	*Oxychilus alliarius*	43 **43**
Clausilia bidentata	38 **39**	*Oxychilus cellarius*	44 **44**
Clausilia dubia	38 **39**	*Oxychilus draparnaldi*	43 **43**
Clausilia rolphii	38 **39**	*Oxychilus helveticus*	44 **44**
Clausilia rugosa	38 **39**	Oxychilus lucidus	43 **43**
Cochlicella acuta	60 **60**	*Pomatias elegans*	22 **22**
Cochlicopa lubrica	26 **26**	*Punctum pygmaeum*	40 **40**
Cochlicopa lubricella	26 **26**	*Pupilla muscorum*	32 **33**
Columella aspera	28 **28**	*Pyramidula rupestris*	27 **27**
Columella edentula	28 **28**	*Retinella nitidula*	46 **47**
Discus rotundatus	40 **40**	*Retinella pura*	46 **47**
Ena montana	36 **36**	*Retinella radiatula*	46 **47**
Ena obscura	36 **36**	Succinea elegans	24 **25**
Euconulus fulvus	49 **49**	*Succinea oblonga*	24 **25**
Eulota fruticum	49 **49**	*Succinea pfeifferi*	24 **25**
Euparypha pisana	50 **51**	*Succinea putris*	24 **25**
Fruticicola fruticum	49 **49**	*Succinea sarsii*	24 **25**
Helicella caperata	58 **59**	*Theba pisana*	50 **51**
Helicella elegans	60 **60**	Trichia hispida	54 **55**
Helicella gigaxii	58 **59**	Trichia striolata	54 **55**
Helicella heripensis	58 **59**	Trochoidea elegans	60 **60**
Helicella itala	58 **59**	*Truncatellina britannica*	28 **28**
Helicella neglecta	58 **59**	*Truncatellina cylindrica*	28 **28**
Helicella virgata	58 **59**	*Vallonia costata*	34 **35**
Helicigona lapicida	50 **51**	*Vallonia excentrica*	34 **35**
Helicodonta obvoluta	50 **51**	*Vallonia pulchella*	34 **35**
Helix aspersa	52 **53**	*Vertigo alpestris*	30 **31**
Helix hortensis	52 **53**	*Vertigo angustior*	29 **31**
Helix nemoralis	52 **53**	*Vertigo antivertigo*	29 **31**
Helix pomatia	52 **53**	Vertigo genesii	29 **31**

Vertigo geyeri	29	31	*Vitrea diaphana*	42 42
Vertigo lilljeborgi	30	31	*Vitrea subrimata*	42 42
Vertigo moulinsiana	30	31	*Vitrina major*	48 48
Vertigo pusilla	29	51	*Vitrina pellucida*	48 48
Vertigo pygmaea	29	51	*Vitrina pyrenaica*	48 48
Vertigo substriata	29	31	*Zonitoides excavatus*	45 45
Vitrea contracta	41	41	*Zonitoides nitidus*	45 45
Vitrea crystallina	41	41		

Index of Species

Synonyms are shown in roman, the correct generic and specific names in italics. Figures in bold type refer to illustrations.

Abida secale	32 **33**	*Hygromia cinctella*	54 **55**
Acanthinula aculeata	34 **35**	*Hygromia hispida*	54 **55**
Acanthinula lamellata	34 **35**	Hygromia liberta	54 **55**
Acicula fusca	22 **22**	*Hygromia limbata*	54 **55**
Acme fusca	22 **22**	*Hygromia striolata*	54 **55**
Arianta arbustorum	50 **51**	*Hygromia subrufescens*	54 **55**
Ashfordia granulata	56 **57**	*Hygromia subvirescens*	54 **55**
Azeca goodalli	26 **26**	*Lacinaria biplicata*	38 **39**
Balea perversa	38 **39**	*Lauria anglica*	32 **33**
Carychium minimum	23 **23**	*Lauria cylindracea*	32 **33**
Carychium tridentatum	23 **23**	*Marpessa laminata*	38 **39**
Catinella arenaria	24 **25**	*Monacha cantiana*	56 **57**
Cecilioides acicula	37 **37**	*Monacha cartusiana*	56 **57**
Cepaea hortensis	52 **53**	*Monacha granulata*	56 **57**
Cepaea nemoralis	52 **53**	*Oxychilus alliarius*	43 **43**
Clausilia bidentata	38 **39**	*Oxychilus cellarius*	44 **44**
Clausilia dubia	38 **39**	*Oxychilus draparnaldi*	43 **43**
Clausilia rolphii	38 **39**	*Oxychilus helveticus*	44 **44**
Clausilia rugosa	38 **39**	Oxychilus lucidus	43 **43**
Cochlicella acuta	60 **60**	*Pomatias elegans*	22 **22**
Cochlicopa lubrica	26 **26**	*Punctum pygmaeum*	40 **40**
Cochlicopa lubricella	26 **26**	*Pupilla muscorum*	32 **33**
Columella aspera	28 **28**	*Pyramidula rupestris*	27 **27**
Columella edentula	28 **28**	*Retinella nitidula*	46 **47**
Discus rotundatus	40 **40**	*Retinella pura*	46 **47**
Ena montana	36 **36**	*Retinella radiatula*	46 **47**
Ena obscura	36 **36**	Succinea elegans	24 **25**
Euconulus fulvus	49 **49**	*Succinea oblonga*	24 **25**
Eulota fruticum	49 **49**	*Succinea pfeifferi*	24 **25**
Euparypha pisana	50 **51**	*Succinea putris*	24 **25**
Fruticicola fruticum	49 **49**	*Succinea sarsii*	24 **25**
Helicella caperata	58 **59**	*Theba pisana*	50 **51**
Helicella elegans	60 **60**	Trichia hispida	54 **55**
Helicella gigaxii	58 **59**	Trichia striolata	54 **55**
Helicella heripensis	58 **59**	Trochoidea elegans	60 **60**
Helicella itala	58 **59**	*Truncatellina britannica*	28 **28**
Helicella neglecta	58 **59**	*Truncatellina cylindrica*	28 **28**
Helicella virgata	58 **59**	*Vallonia costata*	34 **35**
Helicigona lapicida	50 **51**	*Vallonia excentrica*	34 **35**
Helicodonta obvoluta	50 **51**	*Vallonia pulchella*	34 **35**
Helix aspersa	52 **53**	*Vertigo alpestris*	30 **31**
Helix hortensis	52 **53**	*Vertigo angustior*	29 **31**
Helix nemoralis	52 **53**	*Vertigo antivertigo*	29 **31**
Helix pomatia	52 **53**	*Vertigo genesii*	29 **31**

Vertigo geyeri	29 **31**	*Vitrea diaphana*	42 **42**
Vertigo lilljeborgi	30 **31**	*Vitrea subrimata*	42 **42**
Vertigo moulinsiana	30 **31**	*Vitrina major*	48 **48**
Vertigo pusilla	29 **51**	*Vitrina pellucida*	48 **48**
Vertigo pygmaea	29 **51**	*Vitrina pyrenaica*	48 **48**
Vertigo substriata	29 **31**	*Zonitoides excavatus*	45 **45**
Vitrea contracta	41 **41**	*Zonitoides nitidus*	45 **45**
Vitrea crystallina	41 **41**		